STARS
IN THE DARK:
COAL MINES
OF
SOUTHWESTERN
OREGON

DOW BECKHAM

ARAGO BOOKS

Front cover illustration: *Coal miners at a change in shifts, Beaver Hill Mine. (Whereat Collection)*

Book Design: Jeanne E. Galick

Beckham, Dow, 1910 -
Stars in the Dark: Coal Mines of Southwestern Oregon, A History
Bibliography
Includes index

1. Coal Mining—Oregon. 2. Immigrants. 3. Competition.
4. Transportation. 5. Oregon—Mining, Economy.
6. Future Possibilities for Coal.
I. Title

Library of Congress Catalog Card Number 95-075155
Soft cover ISBN: 0-930998-06-5

Published by Arago Books
1283 North 14th Street — P.O. Box 344
Coos Bay, Oregon 97420 — Lake Oswego, Oregon 97034

This book may be purchased from the publisher. Please include $1.50 for postage and handling.

PROLOGUE

Liberio Gonzalas nudged his friend, Leonard Garagaza, and twirled his finger round and round near his temple and pointed his finger at Thomas Kingsley and said, "El loco hombre Escoces esta cantando," (The crazy Scot is singing.) The handsome young Mexicans both laughed and ducked their heads against the storm.

A cold wintry wind was blowing in off the Pacific whipping the falling rain into stinging barbs as it hit their faces that morning. One of the first storms of the fall was blowing itself out as the weather turned cold. It was October 14, 1921, and Thomas Kingsley was a happy man. He had a melody ringing in his heart and he had to let it out. In his deep Scottish brogue, "Bonny Banks of Loch Lomond," was about as difficult to understand as some of the Spanish or Italian spoken by that crew of coal miners. Albert Menegat saw the two young men teasing the Scotsman and told his brother, Joe, who explained to the other Italians and the crew all laughed.

But Thomas Kingsley was undaunted. His world was bright that day and why not; it had been a long five years since he had arrived in America. He and his wife, Mary Jane, had planned that she and Emily would join him within a few months. Instead the British Empire became involved in action against the Kaiser as all Europe broke out into the flames of World War I. Emily was only a baby but now she was almost eight years old and young Tom who was born a few weeks after he had left

Scotland was now a bright young pupil himself. Already young Tom had completed two years of schooling in Scotland and Mary Jane was making him into an instant American. She had taught him all the states and their capitals. He precociously quoted Shakespeare with the encouragement of his mother.

Scotty, as the crew called Kingsley, was 30 years old and the future beckoned. He would have gloried in the life of his namesake, who among other things met the stringent requirements of the U. S. Marines when he served in the South Pacific during World War II, and how he could mimic his mother's delightful way of saying things, such as her statement to Tom and his wife, Dorothy, when they prepared their visit to Scotland many years later: "Rememberrr, lahd a-when you get thairre, you'll find it's the only countrree on airth where they cannah understahnd each aither."[1]

Scotty enjoyed the good-natured banter of the crew, for he knew that they liked him. His heart was too full to even care about that. Liberio was only 21. His flashing white teeth and eyes full of laughter endeared him to all the men. A few men like Albert Menegat who had five children and Thomas McKelvie with six were more sober, but they too enjoyed the camaraderie.[2]

Thomas Hale, the foreman of the crew, was pleased. Little did he care how much the young Mexicans razzed as long they continued to hold up the reputation of their fellow countrymen as being the best of workers. He knew a happy crew could dig lots of coal. They could set a record this day, he was sure. Hale had been carefully checking the equipment and the men to make sure that everything was in order for the day in the mine. The cars gave a lurch as he signaled to Victor Pete, the hoist operator to lower away. The hoist held almost a mile of cable on its huge drum. Down they went and dropped on a slant for three thousand feet to be thirteen hundred feet below sea level at the work site. Little did the young Mexicans realize that the sandy-haired Scotsman would lose his life that day while trying to save theirs,[3] or that Liberio would be the first of the injured men to die in that day's tragedy.

There is no recorded scenario of the activities of that fateful morning at the Beaver Hill Mine, but there are some who still have vivid memories of the tragedy and excitement of that horrible day and have carried them a lifetime. John Menegat,[4] who was six years old, well remembers the sadness that took over his own family, and Alex McKelvie, in his mid-eighties, recalls watching the rescue crew bring out the injured men while standing at the mine entrance. He was thirteen that year. With the night shift alerted, school dismissed for the day and the whole town gathered near the tunnel opening. Alex also remembered that his father had escaped with only slight injuries during the 1916 explosion in that same mine.[5]

1. Tom Kingsley, Interviews with Dow Beckham, April 20 and September 1, 1992.

2. *Coos Bay Times* (Coos Bay, OR), October 14, 1921.

3. Ibid.

4. John Menegat, Interview with Dow Beckham, August 31, 1992; Alex McKelvie, Interview with Dow Beckham, June 10, 1992.

5. McKelvie, op. cit.

TABLE OF CONTENTS

CALIBAN*
IN THE COAL MINES

God we don"t like to complain;
    We know that the mine is no lark.
But there's the pools from the rain;
    But there's the cold and the dark.

God, you don't know what it is—
    You in your well lighted sky—
Watching the meteors whizz;
    warm, with a sun always by.

God, if you had but the moon
    Stuck in your cap for a lamp,
Even you'd tire of it soon.
    Down in the dark and the damp.

Nothing but the blackness above
    And nothing that moves but the cars.
God, if you wish for our love,
    Fling us a handful of stars!!!

—Louis Untermeyer

**Caliban—A slave in Shakespeare's* The Tempest

ACKNOWLEDGEMENTS

This history is about coal mining in Oregon and specifically the coal mines of Coos County. Beginning with the first shipload of coal in 1854, the coal industry was an important part of the economy in southwestern Oregon for over sixty years. The account has been written by using the memories of a few people who worked in the mines. Some, who remember scenes from their childhood when the mines were still operating, have helped. These and unpublished biographies, a few books touching briefly on the subject, and newspaper microfilms were the main sources of information.

A number of people have made this publication possible.

Knowing that some may be left out inadvertently I would, nevertheless, like to thank the following for cooperation and advice: Karl Alpine, one of the last of Coos County coal miners who began working with his father at Riverton in 1921, and later worked for himself a number of years; Leonard and Matilde Gibbs who had their own mine on South Slough that they operated until 1955; Homer Gilfillan who worked in the Southport and Delmar mines in second decade of the twentieth century and gave a graphic account of his experiences; George Chard and his sister, Dorothy Barton, and Bill Anderson who related their experiences in the Overland Mine near Coaledo.

I am grateful also for the memories of Helen (Bomar) Gorbutt, Alex McKelvie, John Menegat, and Tom Kingsley whose relatives were involved in the tragedies at the Beaver Hill Mine.

Others who have added to the account are Thurston (Sticker) Jacobson, Sam Baer, Eleanor (Menegat) Rice, Alice (Schetter) Alter, Florence Lapp, Lionel Youst, John Gilfillan and his wife Edna (Lemon) Gilfillan, Don Whereat, Jerry Phillips, Dr. Robert Macy, Mike Thoele, Patterson Chaney, A. J. "Pi" Moore, Robert Gould, Dan Lynch, Victor West, George Gebhardt, William McKenna, Esther Phillips, Helen and Bud Barnard, Boyd Stone, Robert Pittam, Richard Kutsch, Allan E. Rumbaugh, Martin L. Callery, Jack Beebe, Ed "Doc" Stevenson and Robert Emmett.

John Flanagan, father of Drs. Jack and Roger Flanagan, wrote a comprehensive biography of his father, Patrick Flanagan, the early pioneer who successfully operated the Newport Mine for many years. Dr. Jack Flanagan let me use the story.

Following Patrick Flanagan and the first pioneers came others who were specifically interested in the coal mines. Some were the Monroes, Archers and Brynns. Carol (Archer) Fox has furnished biographical data about her forefathers.

Our Canadian cousins, Warren and Ingrid McWilliams, sent two excellent accounts by Lynne Bowen. Bowen worked with the Coal Tyee Society recording graphic details of the coal mines on Vancouver Island and provided valuable information about the mines and laborers extracting coal.

George Baxter Case did a tremendous amount of research for his master's thesis on the history of the port of Coos Bay from 1852 to 1952. Case graciously allowed me to cite from this well-written account.

I am grateful for the cooperation of the staff of the Coos Bay Library and microfilms of the *Coquille Sentinel*, *Coos Bay Harbor*, *Coos Bay Times*, and *The World*. I thank the staff of Coos County Educational Service District for furnishing data on the Beaver Hill School for the years 1916 and 1921. I am also indebted to the Oregon State Department of Geology and Mineral Industries for the comprehensive survey in 1943-44 by Dr. John Eliot Allen and Dr. Ewart M. Baldwin, and the 1939 "Progress Report on Coos Bay Coal Field" by F. W. Libbey, the director of that department. The Coos County Surveyor helped locate some of the mines. Dr. Douglas V. Keller of Syracuse University permitted me to cite his story of coal fuel development and gave information on the latest chemical experiments.

The compilation of a book sometimes takes on a disproportionate importance in the mind of the creator. Therefore, I would like to thank my wife, Anna, for keeping me in proper perspective, and her patience with me. Anna's memory of many early families in the area and their stories helped on the research. Her proofreading pointed out spelling errors and awkward sentences. I thank my son, David, for his encouragement. I especially thank my son, Mark, and his wife, Cindy, for giving me a personal computer on my 82nd birthday, as well as a series of instructions to get me going. With some basic instruction and several long distance telephone calls from Coos Bay to Portland they soon had me composing texts using regular, bold and italic type, inserting, dragging, deleting, saving, recording endnotes, and the general operation for putting this all together. Mark showed great patience over the telephone while I tried to think in "computer language" and follow his instructions. Also I thank my son, Stephen Dow, without whose help in editing the account and giving advice regarding the research and structure I never could have completed this work.

Several have shared photographs for this volume. I have drawn on the Jack Slattery Collection at the Coos County Historical Museum, the holdings of the Bandon Historical Society, and the files of the *Coquille Sentinel*. Individuals loaning photos have included Lionel Youst, Alex McKelvie, Carol Fox, George Chard, Dorothy Barton, Don Whereat, Esther (McGee) Phillips, and Stephen Dow Beckham. I have also drawn on my own photographs, having lived in Coos County since 1921.

Dow Beckham
Coos Bay, Oregon

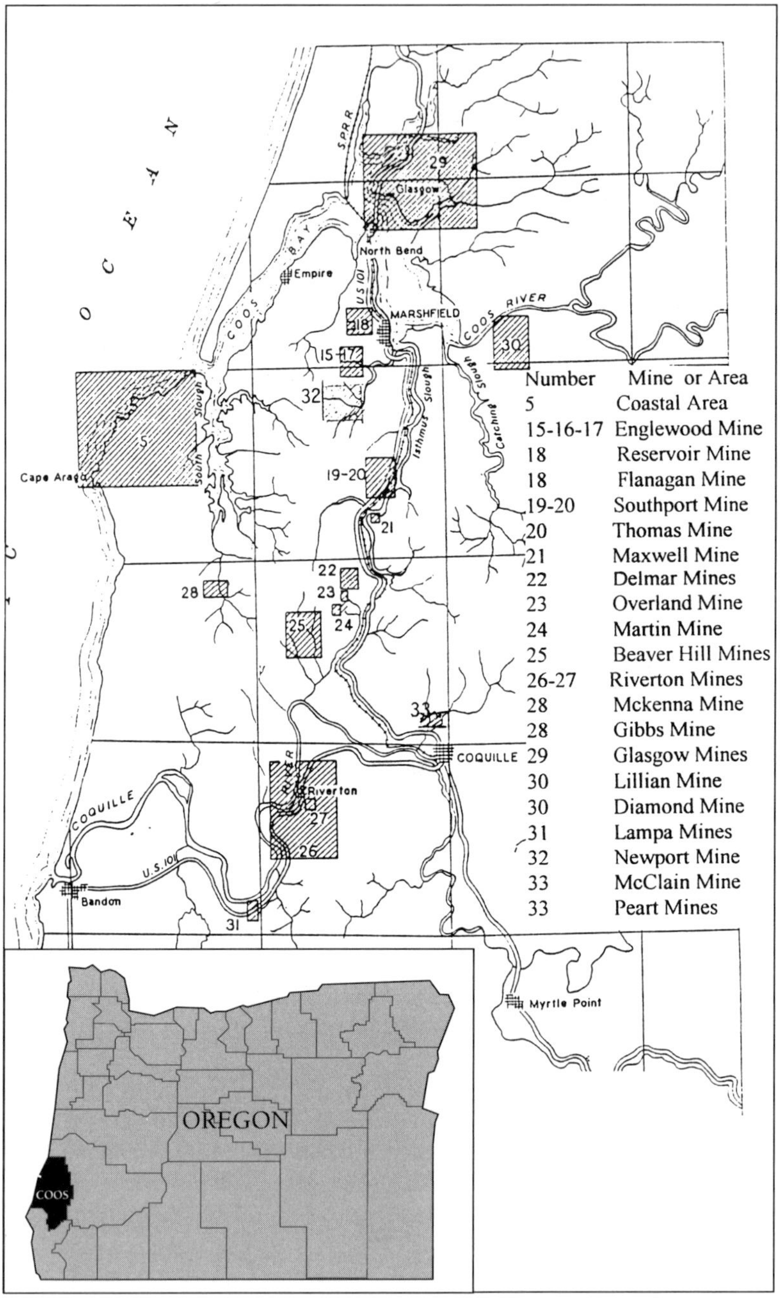

The principal coal mines of the Coos Bay Quadrangle confirmed the size of deposits and their ambitious development. (Allen and Baldwin 1944)

ONE

THE LIBBY MINES

Pioneers discovered coal in the Pacific Northwest as early 1848 in the Cowlitz Valley. The Indians on Vancouver Island had known about the coal there for hundreds of years. By July, 1851, members of William Tichenor's settlement party at Port Orford also located coal veins yielding material that "burned well." The mining and exporting of coal became one of the early successes in British Columbia and in the territory of Washington. The first settlers on Coos Bay also located several coal veins soon after arriving on the bay in the summer of 1853.[1]

The most successful entrepreneur in the Coos Bay region was Patrick Flanagan, who first saw Coos Bay in 1850. A native of Ireland, he was born at Banbridge, County Down, October 23, 1827. His father was in the linen business in Belfast but died when Patrick was thirteen years of age. Soon afterward, the future Coos County financier went to sea as an apprentice seaman. After an eventful period of roaming the world he found his way to Oregon. Patrick's adventures included shipwreck on a desert island, capture by Indians and rounding the Horn many times in early windjammers.[2]

In 1849 First Mate Patrick Flanagan sailed into New York harbor. He was twenty-two years old. He immediately took passage to San Francisco. Like thousands of others, he spent the first year mining for gold. Soon, however, he found a way of making money in the lightering business on San Francisco Bay. Patrick and his brother, James, obtained contracts for loading goods on shallow draft scows from the sailing vessels anchored in the harbor and lightering it to the shore. There were no docks in San Francisco at that time. In later years Patrick indicated that he thought he would have made a fortune if he had stayed in business in San Francisco. However, being of an adventurous nature, he and his brother formed a company with Heman Winchester, John C. T. Hopkins, Horace Paine, Galen Burdett and Eldridge C. Hall.

They intended to explore the possibilities of setting up a trading post at the mouth of the Klamath River in northern California. They planned to freight goods to the mines up river and into the Jacksonville area of southern Oregon. The new company issued stocks at $100 per share to an additional twenty-two investors. The year was 1850, and they sailed north with Captain Lyman on the *Samuel Roberts*. Among the adventurers who took the trips were Rufus Coffin, R. W. Larett, L. Fletcher, Peter Mackey, Dr. Joseph Drew, Dr. Eugene Fiske, Capt. Toney of the barque *Minerva*, John Sawyer, John Hatfield, S. F. Chadwick, A. C. Gibbs, Col. J. Drew, Mr. Gilbert, Benjamin Bratton, Samuel E. Smith, John M. Powell, James McDonald, Jean Baptiste Garnier, Socrates Scholfield, and Patrick and James Flanagan.[3]

The log of Captain William Tichenor of Port Orford provided the following data:

> *On the third day of April 1850 Capt. Tichenor, master of the* J. S. Ryerson, *was laying at anchor off Humbolt Bay. The U. S. schooner,* Ewing, *then on her voyage of recognizance of the coast came to anchor not far distant from the* Ryerson, *communicating with each other. The brig,* Arab, *was in sight standing off the entrance of the bay. The* Samuel Roberts *hailed the* Ryerson *and sent a boat for Capt.Tichenor and took him aboard the* Sam Roberts *as Capt. Tichenor had just made a chart of the coast north and the captain of the* Sam Roberts *wanted all the information he could get of the passage north, which was gladly given. On board the* Sam Roberts *were others that, like the captain, wanted to know all about the coast line north. Those that the captain met were S.F. Chadwick, A.S. Gibbs, James*

and Patrick Flanagan, D. H. Winchester, Col. J. Drew and his brother. The first named became governors of Oregon; Gibbs became prominent in the history of Coos Bay.

The ship missed the mouth of the Klamath River and ended up at the Rogue. There, the captain sent in a small boat with a few men to see if they could enter the river. The boat capsized in the surf. Two men drowned but the rest gained the shore. A large group of Indians captured them. Flanagan had held tightly to an oar and easily made his way through the surf. In a short time the Indians robbed them of all their clothing. Patrick lost his watch to the Indians who took it apart. They tied the pieces about their necks on small cords of buckskin. The men feared for their lives, but the captain bravely sailed over the bar at high tide and fired a cannon. The loud noise of the cannon frightened the Indians who scurried to the timber away from the beach.

After the rescue of the men, the ship continued north along the coast. They failed to see Coos Bay but sailed into the mouth of the Umpqua. Fortunately, they met Levi Scott near the entrance. Scott had previously set up a trading post at the head of tidewater and was there exploring the river. Two men, a Mr. Bullen and a Mr. Sloan from Elkton, accompanied Scott. They were also looking for the possibilities of a harbor on the Umpqua.[4] Captain Lyman took the ship up river, but about half-way up the ship stranded on a sand bar. While awaiting the rising of the tide, some of the members indulged too freely of brandy so the captain had all the brandy cast overboard. Since that date the area has been known as Brandy Bar.[5]

Hearing of the trail from Scott's settlement through the mountains to Jacksonville, Patrick Flanagan left the ship. He arranged with Captain Lyman to have a shipment of goods sent from San Francisco. The *Samuel Roberts* returned to San Francisco in September, 1850.[6]

Flanagan erected a warehouse in Scottsburg on the Umpqua and purchased mules for a pack train. That fall some Indians told of a ship in Coos Bay. Some pioneers called it "Kowes" Bay. Flanagan took Samuel E. Smith with him and they rowed down the Umpqua to the mouth. They then walked down the beach until about opposite the head of North Slough. They went over

Patrick Flanagan sailed to coastal Oregon in 1850 with Winchester, Paine & Company. His energies and partnership with Samuel Mann developed the Newport Mine at the head of Coal Bank Slough. (Beckham Collection)

to the slough and hired an Indian to take them to the bay in his canoe. When they arrived at the bay, they found a ship anchored near where the present railroad bridge crosses Coos Bay. Dozens of Indians surrounded the ship in their canoes. The sailors on the ship kept the Indians at a distance with their muskets. When Flanagan arrived, some of the Indians set up a yell, "Hairy Face, Hairy Face." The Indians on the Umpqua had named him because of his heavy black beard. As Flanagan's canoe neared the ship, the captain kept motioning it back.

Flanagan stood up in the canoe and hailed the captain who in return exclaimed, "My God a white man."[7] Flanagan went aboard the ship, sailed her down the bay, out over the bar, and up the coast to the Umpqua.

With plenty of goods stored in his depot at Scottsburg, Flanagan left with his mule pack train for the Jacksonville mines. It was a long, weary trip to California and back, but the train made regular trips. A couple of years later, in Jacksonville, Perry B. Marple called a meeting of the miners. Marple meant to promote interest in the Coos Bay country. Shortly after the meeting started someone in the crowd yelled, "Here's a man who has been on the bay and can tell us about the coast country."[8]

Flanagan stood upon the platform and told what he knew about the area.[9] Following the meeting, Marple organized a group called the Coos Bay Commercial Company. Perry Marple charged $250 apiece to guide the would-be settlers to the coast. They turned west near the present site of Dillard. When they reached the headwaters of the Coquille River Middle Fork all but nineteen gave up and headed back to Jacksonville. The rest of the company struggled to the mouth of the Coquille River and then traversed the rugged Seven Devil Mountains to Coos Bay.

They chose a site, staked their claims, and started building their cabins. They named the new settlement Empire City. These nineteen were Perry B. Marple, Freeman G. Lockhart, William Harris, R. S. Belknap, Solomon Bowermaster, A. P. Gaskell, C. W. Johnson, M. M. Learn, J. A. J. McVay, Dr. A. B. Overbeck, Charles Pearce, David Rohrer, A. P. Decuis, H. A. Stark, S. K. Temple, Alex Thrift, George L. Weeks, and Samuel Moore. Within weeks J. H. Foster, Jesse Roberts, J. C. Tolman, Dr. V.W. Coffin, Curtis Noble, and William H. Jackson joined them.[10]

Within months others joined these early pioneers. B. F. Ross, having followed the group from Jacksonville, built the first hotel in the new city of Empire. Others who settled were James, Andrew, and John Aiken, John and Mart Davis, Vet Parker, Fortnight (Henry) Miller, David Betchet, James Rooke and John Yoakum.[11]

That fall Freeman Lockhart went back to his cabin in Canyonville for his wife and two children. Freeman's wife, Esther, had been making a living baking and selling pies to travelers while he prospected for gold. From Canyonville, they rode mules and horses along the Umpqua River to the coast. They traveled by the settlements of Elkton and Scottsburg. Dr. A. B. Overbeck, his wife and two little children, and Judge J. C. Tolman, with his family of two, accompanied Freeman and Esther with their two little girls. Esther told of the strenuous effort of riding side-saddle holding her baby in her arms. Her two-year-old child rode behind holding to a kerchief tied about the mother's waist. At Scottsburg, they boarded a scow with their mules and horses. Down the river they went to the beach on the south side. There, they again mounted the animals for the trip down the coast to Coos Bay.[12]

Indians ferried these emigrants across the bay in canoes and the animals swam the expanse. Thus, the first white women came into the area to be later known as Coos County, Oregon. The year was 1853. The Coos Bay Company assigned Lockhart a section on what is now Simpson Heights in North Bend. Lockhart took his small family to the claim where he had previously built a log cabin. By May of 1854, Lockhart opened a coal mine. He sold 30 tons of coal to the captain of a sailing vessel, the *Cynosure*.[13] The captain received $40 per ton for the coal in San Francisco, according to a report. Another report stated that the first shipment of coal out of the bay was on the *Chansey*. That vessel, however, sank on the Coos Bay bar in 1854.[14]

In 1853 some prospectors found gold at Whiskey Run Creek, a small stream entering the ocean about four miles north of the Coquille River. When Patrick Flanagan heard of this he took his forty-mule team south to Coos Bay. Flanagan set up a store at Empire and placed his brother, James, in charge. He loaded his pack train and went south to the head of Isthmus Slough. After spending the night there, he wended his way over the hills to the coast and the "diggins." Gold fever was running high. Miners said they were making fifteen dollars to the pan from sands on the beach. Some rumors were unbelievable, such as the McNamara brothers paying $20,000 for a claim and taking out over $100,000. News spread and soon an estimated 1,000 men gathered there. They staked claims up and down the beach from Whiskey Run Creek. Randolph, as the boom town was called, sprang up like magic. Tents, cabin, restaurants, lodging houses, stores, saloons, made the area into a town.

Winter soon came with a rough surf that prevented any mining. The men contented themselves whipsawing lumber and other activities as they waited out the stormy months. When summer came, the beach did not pan out as expected. Among the hundreds of claims only a few were worth working. The people slowly departed and Randolph became Coos County's first ghost town almost as fast as it boomed into existence. A few years later a fire razed the town with no buildings left.[15] Now-a-days, a beachcomber may find a few broken pieces of glass or pottery on the windswept hardpan where the village stood. Sometimes he or she may find a coin sitting high atop a two-inch sculptured bit of sand.[16]

Within months after the Whiskey Run strike, other miners found gold on Johnson Creek on the Coquille River South Fork. Patrick Flanagan lost no time in loading up his pack mules and heading for the new "diggins" where he set up a store. He freighted down the coast to the Sixes River and over the mountains to Johnson Creek. Flanagan pastured his mules on a large meadow near the head of the Sixes River. The area is still known as Flanagan Prairie.

NEWPORT MINE

From Flanagan Bar on Johnson Creek, Patrick mined $20,000 in gold. He also had a store on the upper Sixes where miners were finding gold. While engaged in business at Randolph, Flanagan heard of the coal discovery on Coos Bay. In 1854 A. G. (Glen) Aiken and his brothers shipped 200 tons of coal. Aiken said,

I came out with my two brothers in 1854. We spent the winter getting out coal from Jim's mine on Boatman's Gulch. We got out 200 tons and Flanagan & Rogers Shipped it to San Francisco for us.

So Flanagan, ever the entrepreneur, got into the coal business early. The freight from Coos Bay to San Francisco was $13 per ton, but the coal brought $40 per ton. Stories vary regarding

The Flanagan and Mann Company not only owned the Newport Mine but also the store and post office at Libby. The water-filled barrels on the ridge of the roof provided fire protection. The beer kegs in the foreground confirmed the miners' thirst. (Slattery Collection)

who opened the Newport Mine at Libby, but Patrick Flanagan and Amos Emerson Rogers bought the mine from the Aiken brothers in 1855.[17] Some of the time it was called the Libby Mine. The Eastport Mine about a mile to the northeast was also called the Libby Mine.

The Newport Mine opened on the south side of Boatman Gulch. The Newport Coal Basin is named from its principal mine, the Newport Mine. The length of the basin north and south from Yoakum Hill to Marshfield (Coos Bay) is about three miles. Coal in this basin included the Reservoir and Flanagan mines. These two mines provided fuel to the water company for pumping. Geologists suggested that the Newport bed probably extended under the town of North Bend. Miners worked two outcroppings there; one was on Simpson Heights and another mine was across from the present day Weyerhaeuser Company chip pile.[18]

Patrick Flanagan soon bought out Rogers and took in his brother, James, for a partner. When the army disbanded Fort

Miners and their families lived on both sides of the railroad tracks which passed through Libby at the Newport Mine. Many who lived at the mine raised a garden and kept a cow and chickens to lessen dependency on buying at the company store. (Slattery Collection)

Umpqua at the mouth of the Umqua River, it left Samuel Stillman Mann without any customers. Mann had operated a sutler store there for the U. S. Army. Flanagan needed Mann's talent and business acumen so he offered him an interest in the mine to start a company store and to keep records. Thus Mann received a third interest in the operation. These two astute businessmen made it convenient for the miners to trade at the store and soon the money Flanagan paid out in wages found its way back into the company coffers.[19]

Flanagan and Mann encouraged their employees in building houses, and soon the village took on the importance of a small town. When they applied for and were granted a post office, they had to name the town something other than Newport. Oregon already had a post office by that name. They named it Libby after an Indian woman who lived there.[20]

The social life in Libby took on a considerable air of importance when, on July 7, 1859, Patrick married Ellen Jane

The Libby Men's Brass Band in 1893 produced music and pride for the small community at the Newport Mine. Washed and dressed in their best clothing, the band members posed in front of the Libby School. (Slattery Collection)

The coal train with its loaded cars passed through Libby about 1895. The miners resided in small dwellings perched along the ravine. (Slattery Collection)

The Libby School, 1893, confirmed the presence of several dozen families dependent upon the Newport Mine.

Winchester. She was the only daughter of T. D. Winchester. Winchester became a prominent attorney in Empire. Patrick was thirty-five years old and his bride was sixteen. Ellen, born in Detroit, Michigan, came west with her father in 1854. Flanagan took his bride to a new home he had built in Libby. Ellen soon became well known and beloved by the whole bay district. Her charitable acts toward the sick and needy were soon noted and appreciated throughout the community.

The Indians were devoted to her as she constantly administered to their welfare. One time the soldiers were looking for Indians to take north to the reservation. After the Indian uprising in southwestern Oregon 1855-56, all Indians were to live on the reservation. Ellen hid Libby in a flour barrel as she did not wish to let the soldiers take her away. Earlier, Patrick Flanagan had taken the contract for $10,000 to remove the Indians from the the lower Rogue to the Grand Ronde and Siletz reservations. They marched and herded 710 Indians up the beach from the Rogue River to Port Orford and from there, June 21, 1856, sailed for the Columbia River on the steamer *Columbia*. Flanagan hired Captain William Tichenor to assist in the endeavor. So these two and an army guard mounted the coastal Indians removal for the federal government.[21]

Patrick and Ellen Flanagan became the parents of seven children: Mrs. T. R. Sheridan, Edward G. Flanagan, Mrs. T. H. Barry, James Flanagan, Mrs. C. F. McCollum, Mrs. H. V. Moffat, and John W. Flanagan. In 1877 Patrick built an imposing twelve room house on a hill overlooking the village of Libby. It was the first plastered house in southwestern Oregon. Flanagan brought carpenters from San Francisco to complete it. The house had a lamp room, bathroom, laundry and six fireplaces. A large, covered porch extended across the front. The home became the meeting place of many pioneer gatherings. Important visitors to the area were always welcome and were asked to stay at the Flanagans'.[22]

The Catholic clergy used Flanagan's Libby home for many years to hold religious services. Circuit riders and other missionaries of different faiths had frequented the area over the years. In August of 1884 the American Baptist built the first church building on the bay. Within a few months of that same year, Trinity Lutherans also completed a church.[23] There was also Neil Monroe who had studied for the ministry but was not an ordained minister. He was religious and felt the call to preach and had quite a following. He baptized many in Coal Bank Slough. There is no record of any sect or congregational connections with Monroe other than the converts became Christians.[24]

Among the people of Coos Bay who worked at Newport during these early days were Ben Roberts, Nels Monson, Thomas Hirst, John Kenyon, Tog Maginnis, David Reese, R. M. Matcheson, Andrew O'Conner, and William Campbell. Dr. C. W. Tower served as mine physician. Other employees included Richard Walters as one of the bookkeepers and William Campbell and Pat Hennessey as underground bosses. Matty Olland was the weighing boss at the tipple. William Archer served as clerk in the store and Al Sleep was the meat cutter. Arthur E. Morton was also a bookkeeper during the years. Another who worked for Flanagan and other mines in the area was Stephen Lapp who came from Germany in 1859. He settled first in the Coquille River Valley but later worked as a logger, farmer and miner. His son, John W. Lapp, also worked in the mines. John W. Lapp was

By the 1890s three portals at the Newport Mine penetrated an estimated 6,000,000 tons of coal in the Newport Basin. Miners posed briefly for the photographer before resuming their labors. (Slattery Collection)

the father of Joseph Lapp and grandfather of Chet Lapp.[25]

The company store at Newport was large and carried a full stock of merchandise. Mann sold sugar, flour, venison, molasses, butter, rice, dried salmon, brown sugar, barrel salmon, salt pork, salt beef, potatoes, tobacco, blankets, clay pipes, knives, candles, tin kettles, horn combs, soap, powder and shot, gunflints, vermillion, beads, brass thimbles, finger rings, thread, ribbon, and cotton, along within numerous other items of necessity for living on the frontier. When a boat arrived at the bunkers down at the bay, Mann hauled tons of freight to the store. J. W. Flanagan in his biography of his father, Patrick, reported that "a number of Italian miners were employed so many barrels of California wine helped make 'steamer day' a lively one."[26]

J. W. Flanagan continued his report:

> *Father ran a large bunkhouse and cookhouse for many years for the single men. George Quigley was the main cook in charge. Milt Elliott was the hired hunter for the cookhouse and it was not an unusual sight to see him coming over the hill with mules laden down with deer, elk and ducks. Cap,*

a Newfoundland dog, was the mail carrier. With a packsack strapped to his back he would start out slowly for the Marshfield post office and the mail. Once the mail was strapped to his back he would head for Libby in a dead run.

People in the community were encouraged to keep their places whitewashed and neat and clean the streets of trash.[27]

Flanagan developed his mine slowly and conservatively and made improvements only as earnings accumulated. At first the coal was conveyed to tidewater by mules on a wooden track. There it was loaded on shallow draft barges. They towed the barges to the ships in the bay. Later, Flanagan extended the road a mile farther down the slough on pilings. They towed the ships backward up the slough to this bunker. The bunker held about 700 tons of coal. He then built a tipple at the mine entrance for weighing, sorting and cleaning the coal. They also added steel strips to the wooden tracks for strength and easier maintenance. All these additions cost money but added to the efficiency of the operation. At times Flanagan had as many men working outside the mine as he did working inside.[28]

Although Flanagan was sharp and frugal and made a lot of his own good luck, he was also a sort of " Prince of Serindip." The mine Flanagan purchased had not previously been surveyed by an experienced engineer, as many investors later learned was absolutely necessary. Yet Flanagan's purchase contained fortunate circumstances. The coal bed rose slightly from the adit. Luckily, and automatically, he avoided the need for costly pumping of water. The vein of coal was almost seven feet in height for easy access by his mules. Additionally, before the tunnel had been extended into the mountain any great distance, the men dug air shafts a short distance to the surface. Later he installed a furnace which created a draft for bringing in good air and exhausting the gases.

Flanagan and Mann made additional improvements within the mine to meet the tipping of the beds and the great distances into the mountain. A visitor in 1874 gave a graphic description of the operation:

You enter the mine tunnel and proceed in a southwesterly direction about a thousand feet before you come to the first

Mule-drawn coal cars stood on the trestle with workers in this view of the Newport Mine at Libby about 1900.

Prior to prohibition of child labor in the Progressive Era at the turn of the twentieth century, boys worked beside their fathers and brothers in the coal mines. A long portal house sheltered the entrance to the Newport Mine about 1895. A ventilation chimney to circulate air rose in the nearby forest above the shaft entrance. (Slattery Collection)

gangway. This gangway runs off in a Northeasterly direction toward the top of the mountain where a furnace is fixed for creating artificial draft and so ventilating the mine. The shafts in Virginia City are all supplied with pure air forced downward, which is unnatural. The tunnels of this mine are ventilated by exhausting the foul air and leaving the pure air to rush in and take its place. All along beside the main tunnel goes an air course which is boarded up tight, so that air can be shut off from portions of the mine where no work is being done. The 'rooms,' as they are called, being chambers from which coal is mined, run east by north, on the alternate sides of the gallery, which is 930 ft in length. Returning over the same ground, we come to the great incline. This is nothing more than an upward pitch of the coal, compelling a tunnel to run through it. They could let the coal drop of its own weight but that would make too much breakage, so they send cars up and down the grade by means of an endless chain rove around a large wooden drum, not vertical but horizontal in its workings.[29]

As funds became available, Flanagan initiated outside improvements. He extended the tramway farther down Coal Bank Slough. The men built this mostly of piling. Later, they filled in with waste from the mines. This is known today as the Red Dike Road. Flanagan cut the *Arcata*, his principal coal shipping vessel, in the middle and lengthened her to forty feet. In the early years of the twentieth century the *Arcata* also hauled coal from the Riverton mines on the Coquille River.[30]

In 1878 Flanagan and Mann bought a small locomotive and built five-ton coal cars in their shops. They paid $4,000 for the locomotive, one of the earliest engines on the south coast. They then handled a greater tonnage of coal with greater dispatch and less breakage. In order to keep abreast with the market, Patrick sent James to San Francisco. James moved his family to the city. In the late 1870s the price of coal rose from $6 to as high as $20 per ton in San Francisco. Flanagan and Mann extended the railway to the main bay, a distance of about two miles. The crews did most of the work along the east side of Coal Bank Slough with wheel barrows and shovels. They cut and leveled

In April, 1875, Flanagan and Mann shipped to Coos Bay a steam locomotive to haul coal from the Newport Mine to bunkers on Coos Bay. (Slattery Collection)

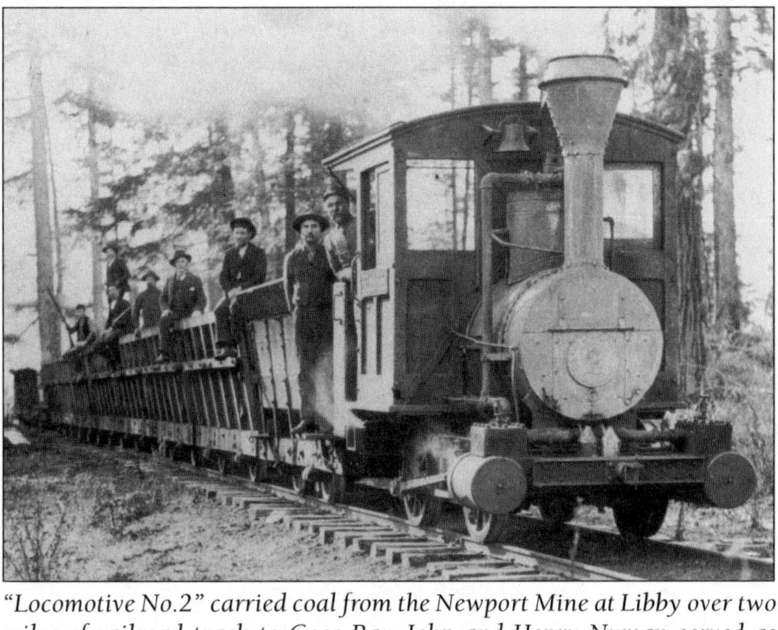

"Locomotive No.2" carried coal from the Newport Mine at Libby over two miles of railroad track to Coos Bay. John and Henry Nyman served as engineer and fireman. (Slattery Collection)

About 1905 a special, low-profile locomotive ran into the shafts at the Newport Mine and hauled coal via Coal Bank Slough to the bunker on upper Coos Bay. Claude Elliott and Conrad Lapp, the boy, appeared in this photograph. (Slattery Collection)

the road bed in the edge of the hills. An item in the *Coos Bay News*, April 19, 1882, read:

> *We have been informed that Flanagan & Mann are making preparations for extending their railroad track to the main channel of Coos Bay near the mouth of Coalbank Slough. The extension will be about a mile and a half in length, and will coat $15,000. The brush has been cleared off portions of the ground, and it is thought that the extension will be finished this summer.*[31]

Another news item of November 22, 1882 stated:

> *Patrick Flanagan returned from San Francisco on the* Arcata, *and we are reliably informed that the firm of Flanagan & Mann will extend the road from their mine to the bay, a short distance south of this place and close to the*

entrance of Coal bank slough. The work will be commenced immediately with the view of completing the road as soon as possible. The extension will be about a mile and half in length, with a gradual incline to the shipping point, so that the coal can be screened while passing through the chutes to the vessel. The proposed improvement will cost between twenty-five and thirty-five thousand dollars, and when completed will be a substantial structure. It is the intention to ship coal steadily from the mine, and with the proposed improvement the proprietors can deliver coal at San Francisco for as small a sum as any mine on the coast.

Flanagan and Mann's crews built a considerable amount of the distance on piling spanning the marshes. Nearby groves of Port Orford White Cedar provided timbers for the trestle. The timbers lasted many years after the mining ceased and local residents used them for firewood. Pete Peterson ran the construction works. Years later, in 1952, his son, Emil Peterson, published the book, *A Century of Coos and Curry*. During the construction, Flanagan's payroll costs climbed to $700 per day, yet the improvements proved worth it.[32] The bunkers on Coos Bay held 800 to 1,000 tons of coal. By 1880, Flanagan and Mann employed 350 men and boys in their operations. They produced as much as 400 tons a day.[33]

In 1883, ever alert to business conditions and especially the fickleness of the coal market, Flanagan and Mann sold out their interests to Captain George Holt. Holt thought he could see possibilities of increasing the production for bigger and better profits. Samuel Mann sold for $85,000.[34] Patrick had bought out the widow of his brother, James, and he owned a larger share. He sold for $100,000, but retained $15,000 in the new company, He also stayed on as manager. Captain Holt came from New England. He was captain on the *Arcata* that had been hauling coal to San Francisco.[35]

Holt was a man of force and a persuasive talker, but somewhat visionary. He organized the company under a new name, the Newport Coal Company. Hawley and Peters of San Francisco invested in the venture. The mine was producing from 30,000 to 40,000 tons of coal a year. Holt had visions of a larger turnover. He built a 1,000 ton bunker at the end of the road to take the place of the loading chutes. Holt hired John Kruse to build the new bunker. Kruse was shipbuilder and superintendent of Asa Simpson's shipyards on Coos Bay.[36]

Holt built the first steel ship on the Pacific Coast specifically to haul coal from Coos Bay to the California markets. He had the ship built in San Francisco and named her *Arago*. Captain Adam Donaldson, born in Ireland in 1867, a sailor who received his captain's papers at age 21, captained the *Arago* for a number of years. He later moved to his farm on Catching Inlet.[37] The *Arago* was a schooner-rigged vessel of 948.19 tons gross. The ship served the coal trade out of Coos Bay until October 20, 1896. That day she sank near the outer end of the north jetty just north of the Coos Bay bar.[38]

Holt subsidized the Eastport mines and the Southport mines by paying their owners a $1,000 a month to cease operation. Patrick Flanagan did not agree with Holt on that cost. Within a

Coal cars at the Newport Mine passed over scales at the top of the bunker. Here the workmen emptied the coal where it spilled by gravity into the hoppers for reloading and movement by railroad to the bunker on Coos Bay. (Slattery Collection)

year and a half after Holt took over, the San Francisco market became flooded with large quantities of coal. A lot of it came from British Columbia. Some came in sailing ships from Eng land and Australia using coal for ballast. The mines in British Columbia considered the San Francisco Bay area their principal market and had been the main competition through the years. Holt had borrowed $40,000 from financier Nicholas Lunning of San Francisco. Since Holt could not meet the payments, Lunning foreclosed and took over the mine and the ships.[39]

The following year, the reverse happened in the market and the price went sky high. Lunning netted $300,000 within that year. He then sold out to Goodall & Perkins, who called the venture Coos Bay Coal and Navigation Company. In 1888, Patrick Flanagan resigned as superintendent shortly after Goodall and Perkins took over the operation. Flanagan moved to Empire where he built another imposing home. In 1889 Patrick Flanagan and Joseph William Bennett established Coos County's first bank in Marshfield. They capitalized the bank at $100,000.[40]

Patrick Flanagan died January 6, 1896. He was visiting his daughters, Mrs. T. H. Barry and Mrs. H. V. Moffat, in San Francisco. Also, his son, John Flanagan, was attending St. Mary's College there.[41]

Samuel Stillman Mann was born in Massachusetts in 1819. In 1849 Mann joined a company on the ship, *Lenore*, and sailed for the gold fields in California. In 1850, along with a number of other settlers, he chartered the bark, *Katie Heath*, and sailed to the Umpqua River. For a short time he operated a store for the soldiers stationed at Fort Umpqua. The fort lay across the river from Winchester Bay. In 1861, when the army abandoned the fort, Mann moved to Empire City and became a partner of Patrick Flanagan in the Newport Mine.[42]

Mann, later at Libby, hired assistants as clerks and meatcutters as well as bookkeepers so that he could follow some personal pursuits. He became Coos County's first judge. In 1865, he married Elphol Tower, daughter of Dr. C. W. Tower. They had two sons, Charles and Fred A. Mann. Mann was reelected county judge in 1866. He also served as U. S. Marshal and collector of customs for a time. In later years he moved to San Francisco. Samuel S. Mann wrote a number of historical essays before his death in 1885. One of his pamphlets, *The Settlement and Early Settlers of Coos Bay*, published anonymously by the *Coast Mail* in 1879, became the source of Hubert Howe Bancroft's history of the Coos Bay Area.[43]

THE EASTPORT MINE

Nelson Northrop started a store in Empire City but soon became convinced that coal was the single product which would lead to financial success. His partner, Simmonds, sold their store in Portland and moved to Coos Bay. They opened a coal outcropping and called it the Eastport Mine. The mine lay about a half mile west of Coal Bank Slough near the later Englewood School. Northrop and Simmons followed Flanagan by about a year (1855) in shipping out their first load of coal. On July 2, 1856, John Evans, United States geologist, bound for the reconnaissance of the Coast Range between Port Orford and the Oregon-California Trail, examined the Eastport mine and wrote in his diary:

> *The coal mine of Mr. Simmonds is situated in a ravine about 3/4 of a mile from the slough, an arm of the bay. It is navigable for ships drawing fifteen feet to this coal mine, or rather to Northrop & Simmonds depot at the termination of a railroad three quarters of a mile from the mine. The main shaft has penetrated three hundred feet, with chambers on each side, but owing to spontaneous combustion labor has been suspended recently and a new shaft opened which has penetrated one hundred feet.*[44]

Until 1874 colliers loaded most of the coal on the ships off barges with shovels. Operators hauled the coal to the barges in wagons and small carts. They towed the barges up the slough, loaded them, and then towed them to the ships. In 1874 Northrop and Simmons built bunkers on Coal Bank Slough. Ships then loaded out in a matter of a few hours by dumping the coal directly into the hold of the ships. This saved three days over the old method of loading. Small ships like the *Arcata* navigated the slough by being towed backward to the bunker. John Parker

Ships ascended Coal Bank Slough to the bunker built in 1874 at Eastport. The mine office (left) and Coal Bank Store stood near the bunker. (Slattery Collection)

A closer look at the store, offices, and other buildings of the Eastport Mine that stood in 1875 on the west side of the Coal Bank Slough about a mile from Coos Bay. This site later became known as Englewood. (Slattery Collection)

supervised the building of the Eastport Mine bunkers and James Aiken furnished the piling. Fred Schetter, assisted by Dick Walters, kept the Coal Bank store established nearby.[45]

Over the years the Eastport Mine continued to disgorge coal for its owners. In the late 1860s Northrup and Simmonds sold out to Charles and John Pershbaker. When these latter partners shifted to lumbering in 1868, they sold the mine for $125,000 to J. L. Pool, Levi Straus, I. W. Raymond, J. S. Dean, Oliver Eldridge, Claus Spreckels, and W. H. Sharp—California capitalists.[46]

The operators opened the tunnel near the section line between the NW1/4 section 3 and the NE1/4 section 4, T26S, R13W. This entrance was at a point about 600 feet south of the section corner and approximately 180 feet elevation. In 1943-44, geologists made a comprehensive study of the Coos County coal beds and reported that the Eastport Mine was a part of the Newport coal bed.[47]

An old tunnel of the Eastport Mine ran north eighty degrees west, sloping about fourteen degrees to the dip of the coal. Gangways ran north and south from the tunnel for a total of a little more than 2,000 feet. Later, the miners extended this slope another 700 feet up the west side of the basin. The owners operated the mine until 1881 and intermittently in later years.[48]

The Libby and Eastport mines on Coalbank Slough confirmed the promise of southwestern Oregon's coal. First in development, these mines prospered where dozens others failed. Their success inspired others to seek "black gold" beneath the forested hills surrounding Coos Bay.

ENDNOTES

1. Stephen Dow Beckham, Coos Bay: *The Pioneer Period*, p. 16.
2. John W. Flanagan, Life of Patrick Flanagan, MS, University of Oregon, Eugene, Ore., p. 1.
3. Ibid., p. 3.
4. Ibid., p. 5.
5. Ibid.
6. Ibid.
7. Ibid., p. 6.
8. Agnes Ruth Sengstacken, *Destination West!*, p. 116.
9. Flanagan, op. cit., p. 6.
10. Orvil Dodge, *Pioneer History of Coos and Curry Counties*, p. 132.
11. Ibid., p. 133.
12. Agnes Ruth Sengstacken, op. cit, p. 113.
13. Stephen Dow Beckham, op. cit., p. 16.
14. Emil Peterson and Alfred Powers, *A Century of Coos and Curry*, p. 391.
15. John W. Flanagan, op. cit, p. 8.
16. I found a half dime at the old Randolph site in the summer of 1940.
17. John W. Flanagan, op. cit. p. 8. Orvil Dodge, op. cit., p. 423: "The Eastport mine was opened at the outset by Northrup and Symmonds, then Charles and John Pershbaker were subsequent proprietors who sold to J. L. Pool. It is now owned by the Southern Oregon Company."
18. John Eliot Allen and Ewart M. Baldwin, *Geology and Coal Resources of the Coos Bay Quadrangle*, Oregon, p. 69.
19. John W. Flanagan, op. cit., p. 12.
20. Ibid., p. 12.
21. Ibid., p. 38.
22. Ibid.
23. Anna Beckham and Dr. Everett W. Curry, *The First Baptist Church of Coos Bay (Marshfield), 1879-1979*, p. 7.
24. Carol Fox, Biography of Forefathers, MS, p. 12.
25. Florence Lapp, Interview with Dow Beckham, September 9, 1992.
26. John W. Flanagan, op. cit., p. 12.
27. Stephen Dow Beckham, op. cit., p. 19.
28. Ibid., p. 21.
29. Stephen Dow Beckham and Rick Minor, Cultural Resource Overview of the Coos Bay BLM District, pp. 163-164. Anonymous newspaper article, October 21, 1874.
30. Bandon Historical Society picture of the *Arcata* loading coal at Riverton.
31. *Coos Bay News*, April 19, 1882.
32. Ibid., p. l64.

33. Stephen Dow Beckham and Rick Minor, op. cit., p. 22.
34. Emil Peterson recorded this as $75,000, op. cit., p. 396.
35. Ibid.
36. Ibid.
37. Emil Peterson and Alfred Powers, op. cit., p. 528.
38. Ernest L. Osborne and Victor West, *Men of Action*, pp. 59-65.
39. Emil Peterson and Alfred Powers, op cit., p. 397.
40. Ibid., pp. 397-398.
41. John W. Flanagan, op. cit., p. 15.
42. Ibid., p. 12.
43. Emil Peterson and Alfred Powers, op. cit., p. 261.
44. Stephen Dow Beckham, op. cit., p. 17.
45. Orvil Dodge, op. cit., pp. 424-425.
46. Stephen Dow Beckham, op. cit., p.17.
47. John Eliot Allen and Ewart M. Baldwin, op. cit., p. 69.
48. Ibid., p. 70.

TWO

MARKETS & COMPETITION

Early day coal mine operators soon found that the San Francisco Bay markets were changeable and usually seasonal in nature. For example, in 1855, Captain Park Butler, in charge of the *S. R. Jackson*, took a cargo of coal from the Eastport Mine. Freight was $10 per ton and, though the coal was as bright and as good as any, Butler had to sell it at $7.50 per ton. The actual cost was $14.00 per ton delivered dockside in San Francisco.[1]

Several conditions and situations affected the market. Customers in the San Francisco Bay area usually stored up coal in the late summer and fall months and then bought sparingly during the winter and spring months. They bought very little during the early summer. Coal producers had to adjust to this erratic market. Sailing ships came into San Francisco by the hundreds carrying coal as ballast from as far away as England and Australia.[2] In 1862 Americans discovered coal near Bellingham Bay and formed the Puget Sound Coal Mining Association. They shipped coal to the California markets.[3] Also, coal came in from Vancouver Island. With the completion of the transcontinental railroad in 1869, the miners in Utah and Wyoming entered the market with a superior grade of coal. Lastly, petroleum and water power gradually took over energy production and ended Coos County coal mining. Natural gas and petroleum production commenced both on the Atlantic Coast and in California in the 1860s. In 1862 developers drilled an oil well in Contra Costa County four miles south of San Pablo.[4] The early industrial development around San Francisco Bay, however, using thousands of tons of coal, continued into the 1910s.

California also holds millions of tons of coal. Beginning in 1861 California miners furnished up to 47 percent of the Bay Area coal needs for over two decades, from the mines on Mt. Diablo. These mines were 40 miles east of San Francisco.[5] The inferior lignite coal failed to compete with products from Oregon, Washington, and British Columbia. After mining the coal near the surface, the cost for additional facilities made their efforts unprofitable. Also, San Francisco Bay area's rapid industrial growth required much more coal and a better grade than the California mines could provide.

San Francisco imported 77,635 tons of coal in 1860, but by 1893, imported 1,479,785 tons. This nineteen-fold increase in demand in a thirty-three-year period reflected industrial growth in the area.[6] Much larger amounts of coal from the Pacific Northwest, especially from the mines on Vancouver Island, met the increased demand. Raymond Sullivan and John Waters, writing on the history of coal and the mines on Mount Diablo, penned these words in the March, 1980, issue of *California Geology*:

> *Because of the limited coal resources in California, large amounts were also imported from Pennsylvania, Great Britain, Australia, and Chile to meet the demand. While foreign coal was higher priced than the local product, this was partially offset by its better quality. The local coal was lignitic or subbitumiuous, while those imported were bituminous and anthracitic. The primary reason for the high price of the imported coal was extra shipping costs which for many years were increased by the lack of return cargo because the West Coast, at this time, had little to export. Ships leaving San Francisco were often forced to sail with ballast on the long voyage around the Horn to the ports of the East Coast, where they were again in the main stream of trade and could load cargo for the remainder of the homeward voyage.*

Right. *The* Breakwater, *its engines fired with coal, passed over the bar and rounded Coos Head to turn south toward San Francisco. (Thomas Kingsley Collection)*

The steamship Arago *moored at the Newport Bunker at the mouth of Coal Bank Slough in this 1890s view.*

> *Beginning in the early part of the 1870's however, the situation changed as California began to produce large amounts of wheat for export, and vessels from abroad no longer had to return empty. This increased revenue made the price of overseas coal more competitive with the poorer quality local product.*[7]

Soon also many other agricultural products such as cattle, fruit and wine entered the economy. Ranchers began to ship millions of bushels of wheat from the Sacramento Valley. Ships then had pay-loads going both ways. One early day rancher, Dr. Hugh Glenn, raised wheat and cattle near Chico in the Sacramento Valley. Glenn was the father-in-law and financial backer of Pete French. French and seven Mexican vaqueros drove 1,200 head of cattle in 1872 through the mountains and high desert of northern California into Harney County, Oregon. Near the Steens Mountain French established his fabulous P Ranch of some 100,000 acres on the Donner and Blitzen River. His cattle drives of 200 miles to the railhead at Winnemucca, Nevada, equaled the famous cattle drives across the Oklahoma Indian Territory.[8]

The Glenn and French ventures in Oregon were representative of California capital in an energetic and often lucrative ex-

Above. *The* Breakwater *took on a cargo at the Beaver Hill Bunker on upper Coos Bay. A popular coastal packet ship, the vessel carried passengers and freight to San Francisco and Portland.*

Right. *Capt. Adam Donaldson served as master of the* Arago *and the* Homer *to haul explosives, coal and lumber in and out of Coos Bay for over 50 years of maritime service in the late nineteenth and early twentieth centuries.*

ploitation of natural resources. What they did to the range, others did to timber and minerals.

Early Coos Bay area miners adjusted to the capricious market. Many of the men who worked in the mines were developing their own homesteads and ranches so they took advantage of the layoffs. They worked at home during the slow season.[9]

In 1860 the Coos Bay mines shipped 3,145 tons of coal to San Francisco. Between 1860 and 1866 Coos County mines supplied the San Francisco market with 2,370 tons a year. This was less than three percent of the market in San Francisco. It was, however, a very important economic factor to coastal southwestern Oregon. Coal production in the Coos

Bay field increased steadily by about 5,000 tons a year. In 1874 the area shipped 44,857 tons of coal to San Francisco Bay. The coal shipments continued to increase with 51,595 tons in 1886 and 69,052 tons in 1890. With the opening of the Beaver Hill Mine in 1896 Coos Bay miners reached their peak year with a total tonnage of 89,960. Production in the fields showed a gradual decrease from that year to only 709 tons going to the San Francisco market in 1915.[10]

The most stringent factor in the California market was the intense competition from the mines on Vancouver Island. The Indians knew of the coal on Vancouver Island long before the British came. Interest in coal increased when the British government decided to change the far-reaching navy from sail to steam-powered ships. Britian had armies in China and India and interests around the globe. If Britian were to continue to rule the waves, she needed coaling stations worldwide.[11]

In 1849 the British Admiralty Office declared Vancouver Island, the land and coal, to be the property of "Her Majesty." The Hudson's Bay Company built a fort at Beaver Harbor to serve as a fur trading post and develop the coal seams. This new establishment was called Fort Rupert in honor of Prince Rupert of the Rhine. Prince Rupert was the dashing soldier who became the first governor of the Hudson's Bay Company in 1670.[12]

At first the Hudson's Bay Company employed Indians to mine coal. Without adequate equipment and mining techniques, however, this proved unsatisfactory. The Indians produced very little coal. The authorities concluded that they should have experienced miners and looked to Scotland as their choice for the best miners. This did not solve the problem quite as easily as they thought it should. Part of the dilemma was that the company officials did not understand the problems: finding proper seams, building outside facilities for handling and shipping coal, constructing houses for miners, and providing the workers with proper tools and equipment. The authorities thought that a miner should take care of all these things. The trained Scotch miners knew how to mine coal in Scotland but the first ones who came to British Columbia simply were not versatile enough to develop a new mine.[13]

The Nann Smith *carried cargoes of "packaged" dimensional lumber from the C.A. Smith Lumber Company sawmill to Bay Point, California. (Slattery Collection)*

Also the attitude of the Scots was part of the problem that no one seemed to recognize. The pride and skill of the Scottish collier was born in serfdom. Seventeenth century legislation bound the Scottish miners like serfs to the coal they worked. Some colliers were bound to a coalmaster for life by accepting "arles," or bounty money. Security for these loans was his children's future labor, thus enslaving another generation. So inevitable was the practice that it became a routine part of the ritual in the baptism of the colliery children.[14]

The government's Emancipation Act of 1775 ended serfdom in Scottish coal fields. The act, however, was full of provisions to prevent the sudden exodus of colliers. It took several years and some petitioning for an individual to gain his freedom. Even then the mining community remained homogeneous and set apart. There were two main reasons for this. Because marriage to a collier meant working in the mine, it was usually necessary for a Scottish miner to marry within the coal fraternity since only colliers' daughters were willing to do such work. Some people criticized the miners for treating mules harshly. They seemed to forget, or never did know, that for generations, women and children carried coal in baskets from the mines. In addition, a Scottish aversion to working underground resulted in a continual scarcity of labor. The miners became increasingly aware of their value and their bargaining power.[15]

By the early nineteenth century, Scottish miners began to realize they had the right to control their own workplace. Industry was booming in the west of Scotland and the owners of the growing number of iron smelters hired colliers from one pit to another. Living conditions improved so much that miners began buying more things for their families. They purchased clocks and furniture for their homes, silver watches and shoes for themselves, silk stockings, and kirk seats for Sunday. Owning these things and their growing interest in education demonstrated their equality to their fellow citizens.[16]

In 1817 a series of sharp drops in the economy threatened this higher standard of living. Some colliers organized unions. The authorities hired thousands of Irishmen as strikebreakers to challenge these unions. Some of the miners were also proud, independent and even arrogant. They thought their skills made them masters of their own destinies.[17]

By the late 1840s Britian started coming out of an economic downturn. The Industrial Revolution began to have its effects. In 1856 Henry Bessemer, an English engineer, developed a converter for producing steel more economically. He revolutionized the process by blasting air into the pig iron. The process oxidized the impurities of silicon and manganese and lowered the cost of making steel.[18]

Jobs were plentiful in the coal fields. Leaving home just to dig coal in far off America was almost unthinkable. Those who did probably would have been unable to explain their motivation. A few, however, accepted the challenge. They seemed overwhelmed with the rugged conditions they found on far away Vancouver Island, where the Kwakiutl Indians considered an enemy's impaled head on a stick an ideal welcoming warning to visitors or newcomers. Such was the response of the Indians near the Nanaimo mines to slave raids of the Haida and other enemies. To European immigrants these sights seemed to confirm their worst fears.

One family that came to this frontier was Andrew Muir, and his father, John, who had read the Hudson's Bay Company advertisement. Muir talked his own sons and nephews into signing a contract for the mines. The Muirs came from Ayrshire, Scotland, where mines had a reputation for producing more coal in a day and being less militant than those from other Scottish counties.[19]

The Hudson's Bay Company designated subsistence money for the miner but not for his family. The authorities told the miners to build their own houses. These first miners lacked the skill to do that. They thought that they had come to mine coal. Wages were forfeited in case of absence. Part of the contract was free passage home unless the miner was dismissed for cause. The company officials' lack of mining experience showed in the contact. Job designations were not in mining terms that the miner could understand.[20]

The regrettable part of the entire situation was that the Hudson's Bay Company had hired expert miners, but they hardly knew how to do anything else. Their expertise as miners consisted of digging coal in a developed mine. They were of little help in building their own houses and the Hudson's Bay Company employees resented the lack of cooperation from the arrogant miners who, after all, had signed up to dig coal. The miners found themselves digging ditches and many other tasks which they resented.[21]

In May of 1850 Muir and McGregor, two of these immigrant miners far from home, lay in a bastion in irons. Their crime was refusing to dig a drain. William Henry McNeill, an irascible ship's captain, ran Fort Rupert like a ship. His word was law. He placed the men in irons and on bread and water for two years. The men felt that they had not committed a crime. They had been there eight months following a six month voyage around Cape Horn. They were disillusioned and fighting mad.[22] These miners did not possess the training for sinking shafts into the mine. There were no proper tools available, long enough drilling rods, nor a pump to remove the water. The situation was further exacerbated by a lack of success in finding a proper seam of coal. Their contract had provided for an annual wage of fifty pounds with a bonus paid for coal dug over the thirty tons. However, until coal was found, it was impossible for the miners to dig for any coal, let alone bonus coal.[23]

The floors of their houses were bare earth covered with broken clamshells. There was only one bake oven in the center of the inner courtyard of the fort for the use of everyone. Other than wild game most food was imported. The Hudson's Bay Company set up a store for its workers and other settlers. Prices were stiff. Each miner dug coal for his account at the store for purchases. It took a lot of coal to pay for some of the items. There were variable price systems. Superior HBC officers paid thirty-three percent over cost; lesser company servants paid fifty to one hundred percent over. The California price of goods regulated the "cash price" to non-company employees. Sometimes there was a 300 percent mark-up.[24] The Indians also resented the whites taking their land and showed resistance in stealing and threatening gestures.[25]

The Hudson's Bay Company leaders continued to try to recruit suitable miners from Europe. They ordered their agent, David Landale, to recruit knowledgeable miners with more versatility. In 1850 he found Boyd Gilmour who seemed to fit the requirements. The company officials did not care to become embroiled in another argument over job designation. Landale still considered the Scots to be the best choice. He said they were a better class of people and accustomed to doing a longer day's work.

By November, 1850, Landale found four miners in Kilmarnock, Ayrshire. Gilmour had a wife and a small family. The Hudson's Bay Company wanted men with small families. Recruitment of men with families was important because it lent stability to the worker who had responsibilities. Also it was more difficult to leave the mine when the miner had to save for moving. One week after his own recruitment, Gilmour wrote to Landale in great distress. Someone had received a letter from Vancouver Island which said the "Muir's party all left the Island and two seamen shot by orders of Mr. Blackendish, who has since been hanged." The news had greatly disturbed the recruits who said that they would rather lie in jail in Scotland than be in irons in Vancouver."

Time was running short for the departure schedule. Landale finally convinced Gilmour that the rumors were only "silly stories." Two of the original recruits withdrew as a result of the rumors, so Gilmour prevailed upon his nephew, Robert Dunsmuir, to accept a position with the party. Versatile and talented, Robert Dunsmuir eventually had a profound effect upon the future of British Columbia and became the leading and most formidable adversary in competition with Oregon coal companies for the California coal market.[26]

The Hudson's Bay Company resorted to a number of means to entice coal miners to the area. A few miners, blacklisted for their union activities in England, came.[27] Hundreds of men, looking for gold, thronged to the Fraser River in the early 1860s. After many of them failed to find gold they went to the Vancouver Island seeking employment. Suddenly the colony had 2,000 inhabitants looking for work.[28] This labor supply was a big improvement for the mine owners over the ten previous years.

Robert Dunsmuir was born in 1825 in a coal region and, in spite of a smattering of schooling acquired at Kilmarnock Academy, was quite uneducated. He seemed to have inherited a shrewd common sense from his Scottish forebears. Dunsmuir and his wife, Joanna (Joan), sailed with Gilmour's party in 1851 on the *Pekin*, a Hudson's Bay Company's 500 ton ship. Joan was pregnant with their first son, James, who was born soon after the party arrived at Fort Vancouver on the Columbia River. She seemed to have stood up under the rigorous sea voyage round Cape Horn. Yet she was not quite prepared for the living conditions she found at Fort Rupert. Perhaps she was determined to hold Robert to his promise of someday building a dream castle like Craigdorroch near where they had lived in Scotland.[29]

In 1855 Dunsmuir completed his contract with the company. He then petitioned the authorities for permission to prospect on his own for coal. For several months he worked under contract for the Hudson's Bay Company and located some lucrative veins near Nanaimo. At Departure Bay, near Nanaimo, he found a substantial vein later known as the Wellington Mines. Dunsmuir found capital in a group of navy officers to develop the mine. They were Lieutenant Wadham N. Diggle of the *H. M. S. Grapler*, Captain Arthur Farquhar, Captain Frederick Wilbraham Egerton, James Harvey, S. H. Rickmann, F. A. Herne, R. Williams, and John Tweedie. Dunsmuir also found additional financing in the firm of Resenfeldt and Bermingham of San Francisco.[30]

At first Dunsmuir's endeavors went under the name of Dunsmuir, Diggle & Company. Soon, however, he bought out his partners and took in his two sons, James and Alexander. The company continued thereafter as Dunsmuir and Sons. By the time of Dunsmuir's discovery the family had moved from Ft. Rupert to Nanaimo. In 1869 Dunsmuir discovered his first important vein. By 1871 he had a large lease on other property and was shipping coal in great abundance to the California markets.[31] Dunsmuir also established mines near the city of Comox, about eighty miles north of Nanaimo. There, he and his sons developed the Cumberland Mines and others that added significantly to their fortunes.[32]

Dunsmuir sent his son, James, to Hamilton Military Academy in Blackburg, Virginia. When he returned at age 25, he had married Laura Surles of North Carolina, a relative of the prominent Byrd family of Virginia. James had previously studied in Portland, Oregon, as an apprentice welder and mechanic. He had been active in the actual mining of the coal in his father's mines. "Mrs. James," as Laura became known, was a talented singer and fit nicely into the social life of Victoria, then a prospering city. Later her husband became premier and lieutenant-governor of the province.[33]

Dunsmuir sent his second son, Alexander, to San Francisco to take over marketing of their product. Alexander became acquainted with some of the leading capitalists on the West Coast. These men, some known as "the bonanza kings," included Charles Crocker, Mark Hopkins, James Fair, and Collis Huntington, the railroad magnate. They built their mansions on Knob Hill and probably inspired Robert Dunsmuir to try to out do them when he finally got around to building a stone castle in Victoria for his wife.

In 1889 Alexander Dunsmuir also built a fabulous mansion on a large estate near San Leandro at a cost of $350,000.[34] The

Joan and Robert Dunsmuir prospered through their coal mines on Vancouver Island. They spent part of their fortune to erect the mansion "Craigdarroch" on a hillside overlooking the city of Victoria. The massive, three story house of dressed stone affirmed that the Dunsmuirs had arrived financially and socially. (BC Archives and Records Service Negs. 95143, 68228 and 23892)

Dunsmuir connections with the California capitalists led to giving their name to the railhead of Pusher, southwest of Mt. Shasta. This became the small city of Dunsmuir, California, where the Southern Pacific had massed its construction crews to extend the rail line north through the Siskiyous into southern Oregon.[35]

Dunsmuir also relied on these California capitalists for funds to build the Esquimalt & Nanaimo Railway. Dunsmuir had become wealthy but he could see the greater advantage of a railroad. He wanted the land grants and mineral rights accompanying the construction of the rail lines to add to his wealth. The land grant, in part, was a reward for having helped Sir John A. MacDonald complete the railway across Canada. The rail line was to lure British Columbia away from its California connections. The completion of the railroad was a promise if the province would join the Canadian confederation. When built, the Canadian Pacific crossed the plains and rugged mountains of the West with a thousand miles more track than the first transcontinental line across the United States.[36]

Joan Dunsmuir proved to be a staunch pioneer and finally saw Robert keep his promise in which he had said:

> *If you will accompany me to America I will build you a castle of stone like Craigdarroch which will have stained-glass windows, a tower, ballroom, servants' quarters and you will have diamond tiaras and liveried coachmen and you can give grand dinner parties under crystal chandeliers.*

Craigdarroch, the Dunsmuir home in Victoria, was estimated at over $500,000 when completed in the 1880s. It stood in a vast estate of woods and formal terraced gardens. The building consisted of four stories elegantly furnished. Above the tower lookout, a round chamber had floor tiles set with mosaics. Robert Dunsmuir kept his promise but died, April 13, 1889, before he was able to live there.[37] Joan, however, lived in the fabulous mansion many years and saw her daughters married in splendor. Two married into European royalty.

Robert Dunsmuir had the reputation for being a hard and unrelenting taskmaster. Many coal miners, especially those who were active members of miners' protective associations, regarded Robert Dunsmuir as a "grasping monopolist and an autocratic tyrant." Residents of Victoria held a different view. To them, he was the city's commercial savior. In March of 1888, when he brought the tracks of the Esquimalt and Nanaimo rail line on through to Victoria, he was greeted as a civic hero. Huge crowds gathered to meet him and applaud his success.[38]

James Dunsmuir more than doubled the coal production when he took over management of the company. At times the mines loaded out as much as 1600 tons per day for the California markets. Unlike his father, James was stiff-necked and short-tempered. He seemed unwilling to put himself out to produce a better working situation. His father was paternalistic and demanding, but he had sympathy for the miners' problems and attempted to improve working conditions as long as the men conformed to what he thought was best.[39]

Some of the worst mining tragedies and strife in British Columbia occurred during the tenure of James Dunsmuir and added to the despotic reputation of the family. Many alleged that greed caused the death of several miners at South Wellington. The mine accidentally flooded February 15, 1915, by a miscalculation in digging a shaft. The final explosion in the digging broke through into a flooded mine that immediately covered the men in the lower section.[40] Another instance, on October 5, 1909, thirty miners at Ladysmith, died in an explosion because Dunsmuir would not provide safety lamps. Open flame lamps used by the miners ignited the mine gas. Ladysmith is a small town a few miles south of Nanaimo.[41]

The Dunsmuir legacy foundered on the lack of heirs in the third generation. By 1910, the coal interests were sold to Canadian colliers, and by the 1930s, most of the fortune had vanished.[42]

The Dunsmuirs competed with southwestern Oregon coal investors and coped with many of the same problems. The Dunsmuir family became far richer and better known than any of the Coos County mine owners, but, in contrast to the situation in British Columbia, many of the Oregon mines were owned by absentee investors such as the wealthy of San Francisco. In that respect the "black gold" of Coos County was

Marshfield, Oregon, looking west toward Blossom Gulch (left) and Mingus Lake (right), August 1912. (Stadden Collection)

but one of their economic ventures and did not confirm special wealth or position as that nurtured by the Dunsmuirs in Canada.

ENDNOTES

1. Emil R. Peterson and Alfred Powers, *A Century of Coos and Curry*, p. 391.
2. Ibid., p. 395.
3. Lynne Bowen, *Three Dollar Dreams*, p. 81.
4. Raymond Sullivan and John Waters, Mount Diablo Coalfield, *California Geology*, March 1980, p. 58.
5. Ibid., pp. 57-58.
6. Ibid., p. 58.
7. Ibid.
8. Giles French, *Cattle County of Peter French*, pp. 41-46.
9. Emil R. Peterson and Alfred Powers, op. cit., p. 394.

10. George Baxter Case, The History of the Port of Coos Bay, 1852-1952, MA Thesis, 1983, p. 104.
11. Lynne Bowen, op. cit., pp. 78-79.
12. Ibid., p. 79.
13. Ibid., p. 24.
14. Ibid.
15. Ibid.
16. Ibid., p. 25
17. Ibid.
18. Lynne Sabel and Philip Steele, *1,000 Great Events*, p. 227.
19. Lynne Bowen, op. cit., p. 15.
20. Ibid., p 28.
21. Ibid.
22. Ibid., p. 32.
23. Ibid., pp. 28-29.
24. Ibid., p. 29.
25. Ibid., pp. 78-79.
26. Ibid., pp. 46-47.
27. Ibid., p. 125.
28. Ibid.
29. James Audain, *From Coalmine to Castle*, p. 112.
30. Lynne Bowen, op. cit., pp. 137-139.
31. Ibid., p. 139.
32. Ibid., p. 315.
33. Ibid., p. 148.
34. *San Francisco Chronicle*, March 8, 1971.
35. James Audain, op. cit., p. 93.
36. R. W. Gosnell, *Sixty Years of Progress*, (A pamphlet abridged by the Kiwanis Club of Victoria, B. C., 1970).
37. Ibid.
38. *The Islander*, October 25, 1992.
39. Lynne Bowen, op. cit., p. 319.
40. Ibid., p. 52..
41. Ibid., p. 30.
42. Ibid., p. 22.

40. Ibid., p. 52..
41. Ibid., p. 30.
42. Ibid., p. 22.

THREE

TRANSPORTATION

Capitalists who ventured investments in the Coos County mines faced a number of risks. Investors confronted competition both locally and from the mines of British Columbia and Washington. They also dealt with labor problems. Should the workers organize and demand higher wages or better working conditions, the marginal mining company might fail in days or weeks. Even well established mines, such as Patrick Flanagan's, had to contend with the possibility of losing tons of coal on the treacherous Coos Bay bar and the storms in the mighty Pacific. In the 1850s and 1860s, square riggers plied the waters of the coast and carried on the trade. They were principally barkentines, brigantines, and some topsail schooners.[1] The misfortunes of sail-powered shipping were a cloud on the prosperity of the southwestern Oregon coal mines.

RIVER AND HARBOR IMPROVEMENTS

Shippers needed harbor improvements, namely a lighthouse, jetties, tug service, life-saving crews, harbor lights, and channel markers. During the late 1850s and 1860s the tug service was little better than none at all. The first tug on the bay was the *Hartford*. Asa Simpson brought her to Coos Bay in 1857, right after he built his North Bend sawmill. The boat was not suitable for the work so Simpson took her back to San Francisco. He removed the engines and placed them in an old brigantine, the

Star of China. The *Star of China* had been an opium trader in the Far East. Simpson changed her name to *Fearless* after converting her into a tug with steam engines. The *Fearless* moored at the North Bend dock until notified that a vessel was off the bar.[2]

Often vessels waited outside the entrance for several days for the tug to appear. The captains grew impatient in waiting. Sometimes they squared away and sailed into the harbor. Before the government completed the jetties in the twentieth century, the bar was much closer to the headland than afterward. The main channel was in the vicinity of Charleston. Where the north jetty now rests on the North Spit, there was a middle ground with a shallow channel north of that. Often, in attempting to sail in, vessels would lose wind under the lea of the headland and strike on the middle ground where there was a heavy break. The disabled ship might then go on the beach somewhere between Rocky Point and Coos Head. At times the tug got a line on a vessel stranded on the "middle ground" and pulled it back into the main channel. All too often, however, the ships were lost. Many vessels, loaded with coal and lumber, met disaster in the treacherous currents and ponderous waves on the way out of the bay.[3]

Early lumbermen on Coos Bay and on the Coquille River established shipyards in conjunction with their sawmills. Captain Asa Simpson located his yard in North Bend. In 1867 he employed John Kruse, a veteran shipbuilder born in the Danish-speaking part of Germany. Kruse came to the United States in 1854. During the thirty years Kruse worked for Simpson he built twenty-eight ships. One of his greatest achievements was the construction of the *Western Shore*, a masterpiece of beauty and speed. Simpson launched her in 1874 at the cost of $70,000. Her sleek lines and towering masts testified to her speed. This clipper ship made a fantastic trip from San Francisco "around the Horn" to Liverpool, England in 103 days. Other men who later built ships for Simpson were Emil and H. E. Heuckendorff, Knud Vladimar Kruse, Victor Anderson, and Peter Loggie.[4]

Aaron Lobree also had a shipyard at his mill near the mouth of Isthmus Slough. Captain Reed, who commanded the *Arago* several years, built ships for Lobree. E. B. Dean Lumber Company in Marshfield operated a small shipyard in the 1880s.[5]

Relative calm surf in 1915 enabled the U.S. Life-Saving Service crews to save those aboard the ill-fated Santa Clara *which grounded at Coos Head. (Stadden Collection)*

Shipbuilders also constructed approximately 100 vessels on the Coquille River between 1871 and 1945.[6] Raine and Johanson built the first sea-going vessel on the Coquille at Grube's small, water-powered sawmill on the lower river. They named her the *Eliza Johanson.*[7] Ernest L. Osborne in *Wooden Ships and Master Craftsmen* stated that the first boat built on the Coquille River was the *Mary*. James C. Fitzgerald and Daniel Giles built her in 1871. They built the *Mary* for the river trade, but the vessel did make a sea voyage to Gold Beach.[8] According to Daniel Giles's reminiscences as quoted by Alice Wooldridge, Giles and a partner built a small boat and launched it at the mouth of the Coquille River in April 1854. They sailed to Port Orford and brought supplies to the miners at Randolph.[9]

All the activity in shipbuilding added no revenue to the coal operators. Most of the ships hauling coal and lumber, as well as agricultural products from the area, also carried passengers. Safety of the passengers, the cargo, and, of course, valuable ships became a serious concern. Vessels lost on the Coos Bay bar in the 1850s and 1860s were the *Quadratus* (three lives lost), *Colianna*, *Chansey*, *Jackson*, *Cyclops*, *Noyd*, *New World*, *D. M. Hall*, *Ida Rogers* and the *Energy*. In 1870 the brig *Devans* and the schooner *Advance* wrecked. The tug *Fearless* foundered October 13, 1873, (one life lost). The *Julia Ray* wrecked January 25, 1880, and the *Charles W. Wetmore* sank near the harbor entrance September 8, 1892. The *Julia Ray* and the *Charles W. Wetmore* were steamships.[10] The steamer *Emily* foundered July 7, 1893. The loss of so many ships and the burgeoning export of coal and lumber proved the need of installing some aids to navigation.[11]

On July 2, 1864, Congress appropriated $15,000 to erect a lighthouse near the entrance of Coos Bay on Gregory Point. The Cape Arago Lighthouse, located on the western terminus of Lighthouse Island, first illuminated its beacon on November 1, 1866. In 1894 the government appropriated $15,000 for improvements to the station. These investments included a new keepers' duplex on the island. The crews also enclosed the wrought iron lighthouse tower with brick. In 1908 the government erected a new, wood frame lighthouse on the eastern edge of the island. The department replaced that building by a third, concrete lighthouse in 1934. The U. S. Coast Guard installed an automatic beacon in the 1950s.[12]

On March 3, 1891, Congress appropriated $50,000 to establish a lighthouse on the north spit of the Coquille River. The station illuminated its beacon in 1896 and remained in active service until 1939.[13] That structure is now a tourist attraction in the Bullards Beach State Park.

In 1878 the federal government established the first U. S. Life-Saving Station at the Cape Arago Lighthouse. A solitary keeper took up quarters in a small, wood frame building in the channel between the Lighthouse Island and the mainland. His

Right. *Designed by Robert Stockton Williamson and fabricated by Anthony Bien in San Francisco, crews erected the original Cape Arago Lighthouse in 1866 at the tip of the island at Gregory Point.*

ability to go to the assistance of seamen in distress was contingent upon volunteers rushing south along the estuary and coast from Empire City to the station to join him in manning the surfboat.[14]

In 1891 the U. S. government activated the station to full status and moved the quarters to the North Spit, approximately two miles north of the harbor entrance. The new station was equipped with shot lines, whip lines, a lifeboat, a surfboat, and breeches buoy. The keeper trained his men to handle boats with oars in the raging surf. They saved ships and many lives through the years. In 1916 the Life-Saving Service moved the quarters to Charleston, nearer the entrance to the harbor. Also, in the 1890s, the government established a life-saving station at the south side of the Coquille River in Bandon, Oregon.[15]

Critical to safe navigation was accurate information for mariners. In June, 1861, the U. S. Coast Survey began the formal charting of Coos Bay. Headed by James S. Lawson on the brig *R. H. Fauntleroy*, the crew completed the survey in 1866. The mapping and hydrographic study provided a sound base for future engineering work in the harbor. Yet, it was almost two decades before improvement activities began in 1879.[16]

Engineer Channing M. Bolton surveyed the harbor entrance in 1878. Bolton recommended building a jetty from Fossil Point across the mouth of South Slough. He blamed the huge eddy caused by the currents of South Slough meeting with the currents from the upper bay as the main source of trouble. The swirling currents kept the natural flow of the tides from scouring out the harbor entrance. At times the sand made the bar so shallow that even light draft sailing schooners could not cross.[17]

In 1878 Congress appropriated $972,000 (later reduced to $600,000) for improving the harbor entrance. In 1879 an advance of $40,000 initiated the work. Bolton's design called for huge log cribs filled with immense stones quarried on Coos River. E. B. Dean and Company contracted to build the cribs and William Luse delivered the rock on barges to the jetty site. The workers placed the first crib April 6, 1880, followed by a series of cribs over the next ten years. This kind of jetty was known as a "training" jetty on the East Coast. Jetties of this type scoured

Steam tugboats moved massive boulders from the quarry above the forks of Coos River to the North Spit in the 1890s for the jetty.

out the sand, thus deepening and shaping the channel by guiding the current. The placement of the huge cribs was difficult in the shallow water, but when the workmen reached a depth of forty feet the progress slowed drastically. They persevered, however, and finally reached a distance of 1,825 feet, still a long way from South Slough. The engineers reconsidered and discontinued the project July 21, 1890.[18]

The Corps of Engineers then appointed a new committee that changed the entire approach. The committee recommended two external jetties 1,500 feet apart extending into the sea. To assume the design of the new project at Coos Bay, the Corps selected engineer James Suydam Polhemus, a graduate of Lehigh University. They hired Polhemus because of his extensive experience in building jetties. He had worked on the Columbia and Yaquina jetties. He also had worked on the Great Lakes and the Gulf of Mexico harbors.[19]

In 1890 with an appropriation of $125,000, Polhemus began the North Jetty. In August, 1891, from their dock on the North Spit, his crews built a tramway for hauling the stones to

A coal-powered steamer waited at the wharf in Millington in 1922 on Isthmus Slough to receive a large cargo of timber. Longshoremen loaded the lumber piece by piece for the deck cargo. Coal went in the hold. (Stadden Collection)

the jetty. By March, 1892, the jetty reached some 1,800 feet into the ocean. The crews sank brush mattresses to help hold the huge boulders. In 1892 Congress appropriated $210,000 so the work could continue. After a series of additional appropriations the North Jetty was finally considered complete on March 15, 1901.[20] Anson Rogers' South Coos River ranch also furnished some of the rock.[21]

The jetty was a remarkable improvement. Before the huge stones began to show above water at low tide, the current scoured the bar for a greater depth. In 1892 the low tide channel depth at the entrance increased from eleven feet to an average of seventeen feet. Later, dredging help to maintain greater depths.[22] Even with the improvements, the harbor entrance continued to take lives and ships. The raging waves and cross currents made navigation of the bar extremely hazardous during inclement weather.

In September, 1890, the Chief of Engineers ordered a preliminary survey of the upper harbor. Captain Thomas W. Symons of the Corps of Engineers found several shoals that impeded traffic except at high tide. Some shoals presented serious problems. To remove these navigational barriers Captain Symons recommended the building of a combination dredge and snag puller for use in Coos River and the sloughs, as well as in the bay. Later the Corps added a pump dredge and piped the material for fills. Gradually, the dredges deepened the channel all the way up the estuary to the mouth of Isthmus Slough. Ultimately, especially for the lower bay, the engineers recommended a hopper dredge. The government subsequently built the *Colonel P. S. Michie* in Seattle. The construction yards launched the dredge August 16, 1913. It arrived at Coos Bay the following January and, for many years, the Corps continued to improve the harbor channel.[23]

Another federal responsibility was the maintenance of harbor lights in the region's estuaries. In 1891 the Lighthouse Board reported on the need for such aids to navigation at Coos Bay:

> *It is found that to enable vessels to cross Coos Bay bar at high water, which are, owing to the tides, often obliged to navigate the bay at night, it is needful to establish certain beacon lights.*

Shortly after publishing this report the Board established range lights on Coos Bay. The keeper constructed and maintained channel markers and lit the kerosene beacons to guide shipping on this estuary.[24]

In 1911 Nils Petter Adamson, former assistant keeper of Desdamona Sands Lighthouse at the mouth of the Columbia River, moved to Coos Bay as harbor light master. Subsequent to serving on the Columbia, Adamson became, in 1906, the first keeper of Eldred Rock Lighthouse in Alaska. When he retired in 1936 his son, Chester Adamson, succeeded him. When the younger Adamson retired in 1976, the Adamsons had completed sixty-five years of continuous range light maintenance on Coos Bay. Chester Adamson was one of the last civilian workers in the U. S. Coast Guard. A crew of six took over the duties.[25]

Other improvements included wireless stations along the coast. In 1907 the government established a wireless station on Cape Blanco. The early telegraph system was primitive; however, it saved lives and provided a great new means of communication from the sea. The receiving poles at Cape

Blanco were 180 feet high. When a message was expected, the operator sat for several hours at the instrument, with tubes from the delicate receiver in his ears. He listened intently for the little tick of the instrument. Ships from Alaska began sending messages as far away as a thousand miles.[26] In July of 1917, the authorities built a new station near Marshfield, and discontinued the receiving station on the cape. They set up the new station just south of Marshfield on Coal Bank Slough. Because of the war in Europe, U. S. Marines guarded the Cape Blanco station. The Marines continued their guard duties at the new station during the war years.[27]

The jetties, dredging, harbor range lights, and wireless communications improved safety for shipping coal, lumber and passengers from Coos Bay. The bar, however, remained a place of danger until the South Jetty was finished in 1928. Not all of the ship strandings on the Coos Bay bar ended in tragedy but often they caused a loss to the owners and shippers. The steamer *Arago*, outward bound with passengers and a load of coal on Saturday, October 2, 1891, met with trouble. The tug *Liberty* was steaming in with a vessel in tow. Captain Adam Donaldson of the *Arago*, while trying to avoid the tug, got too close to the South Spit. The steamer went aground about 250 yards northwest of Guano Rock. On Sunday, the tug *Hooter* shot a line to her but was unable to pull her off the sand. On the Monday afternoon at high tide, however, after Captain Donaldson had jettisoned most of the cargo of coal, the *Hunter* and the *Liberty* towed the ship back into deep water.[28]

Five years later, almost to the day, the *Arago* met with disaster. She foundered on the way to sea, on October 20, 1896, with a great loss of life. The *Arago* was a schooner-rigged vessel of 948.19 tons gross burden, belonging to the Oregon Coal and Transportation Company of San Francisco, where she was constructed in 1885. Her owners valued her at $100,000 at the time of her loss. On the morning of the disaster, she headed seaward from Empire City laden with 750 tons of coal and fifty tons of general merchandise. In addition to her cargo, she had seven passengers and a crew of twenty-five sailors including the officers. Alfred Reed, a shipmaster of long experience, was the cap-

Above. *Nils P. Adamson named the Range Light tender the* Hilma *for his Swedish wife. His son, Chester N. Adamson, the boy seated in the boat, succeeded to the post as tender of the Coos Bay Range Lights and served for 41 years. (Beckham Collection)*

Right. *Nils P. Adamson, born in Sweden in 1869, served in the Life-Saving and Light-House Service at the mouth of the Columbia River and in Southeast Alaska. In 1911 Adamson became keeper of the Coos Bay Range Lights. (Beckham Collection)*

tain. Reed had a reputation of being a thoroughly efficient sailor, as well as an excellent bar pilot, cool and possessed in times of danger.[29]

The weather was hazy and heavy seas lashed the bar. Near the bar she encountered a powerful northerly current that swung her toward the jetty. The helmsman quickly put her hard a-starboard and called for full steam ahead. She swung around and struck the submerged rocks of the jetty. The collision ripped a huge hole in her after compartment. She began to fill and sink by the stern.[30]

The sea was so rough that great rollers instantly swept her decks, carrying death and destruction with them. The breakers soon tore away the after cabin and took with it five of the passengers. All perished in the crash. The strong rush of the current continued to force the disabled vessel along until she finally came to rest about 200 yards north of the jetty. Within a very short time, the hull of the ship sank out of sight, leaving survivors hanging in the rigging. The captain ordered the crew to launch the two lifeboats when he saw that the ship was sinking. He directed the boats to stay as near as possible and attempt to take off additional personnel.[31]

The first mate, in charge of Boat No. 1, handled her very well, keeping near the sunken vessel. Unfortunately, his boat capsized in the rough sea. The men quickly righted the boat but before they could bail the water, she overturned again. The raging sea claimed all of her occupants except the mate and two other men. These three succeeded in swimming to Boat No. 2. Noticing that Boat No. 2 then contained twelve occupants, the captain ordered the mate to leave and attempt to reach the harbor. The first mate took over and made his way around and over the bar to the safety of the government jetty dock.[32]

With no boat nearby and huge waves lashing away at them, those clinging to rigging despaired of their precarious position. Fortunately, Captain Reed had blown his whistle announcing the disaster. He had whistled a number of repeated blasts two or three minutes before the waters inside the ship extinguished the fires. Reed kept calm and assured his men that a lifeboat was nearby and would soon be coming to their assistance.[33]

The tug *Columbia* came down the bay about a quarter of a mile behind the *Arago*. She heard and answered Captain Reed's signals and steamed to the government dock. The government dock was at the tip of North Spit across from Fossil Point.[34] Captain Magee asked the watchman there to notify the Cape Arago Life-Saving crew about the *Arago*. A short time before the government had moved the station from Lighthouse Island to the North Spit, about two miles from the bar. In that storm it would have been impossible to have launched a boat at the lighthouse.[35]

At the new North Spit location Keeper Haynes launched a

surfboat in response to the signal from the *Columbia*. The surfmen rowed down the bay about a mile to the government works. They still could not see the wrecked ship. Keeper Haynes dispatched two men to the north jetty. They took a hand car on the rails to the end of the jetty. They soon saw that a surfboat was about the only chance of saving the men in the rigging. Within five minutes they reported back to Haynes. Still almost two miles away from the wreck, the men strove against the tide with all their might and finally rounded the end of the jetty. Soon they hove in sight of the stranded men. During this time there were eight men clinging to the shrouds of the *Arago*. The surfboat appeared from behind the jetty "fairly flying," as Captain Reed afterward expressed it. Stripped to their waists, the surfmen pulled as if their own lives were at stake.[36]

All around the *Arago's* dangerous wreckage of booms, spars, and timbers hindered the rescue attempt. The swift current forced the surfmen steadily in a northerly direction. Great combers raced upon them from the west. Every moment the lifeboat was in danger of broaching. The crew exerted its utmost strength, while holding her head to the sea and maintaining a position near the wreck. Fighting the seas and the wreckage, the keeper skillfully edged closer to the stricken vessel. After several attempts, he successfully threw a heaving line to the men hanging in the rigging.[37]

As soon as the shipwrecked men got the heaving line, the keeper, Captain Haynes, tied his end of it to the lifeboat's hawser. The men on the wrecked ship pulled the hawser into the rigging. Captain Reed directed a man to fasten himself to the hawser and then jump overboard. The surfmen pulled him through the water to the surfboat. They successfully saved five men in this manner. Soon, however, they noticed that the mast was about to collapse. The keeper ordered the last three, including Captain Reed, to lash themselves to the hawser for the final rescue attempt.[38]

In less than an hour after the life-saving crew had rounded the jetty they had all the remaining survivors in their boat. Once they reached the calmer waters of the lower bay, the *Columbia* took them in tow. Thirteen people out of thirty-two lost their lives in the mishap. Captain Reed wrote a letter of commendation to Major T. J. Blakeney, Superintendent of the Twelfth District, U. S. Life-Saving Service in San Francisco, California, November 1, 1896:

I take pleasure in writing you a few lines regarding the wreck of the steamer Arago, *near Coos Bay jetty on October 20, 1896, and the performance of the life saving crew on that occasion, particularly of Captain Haynes, who was in command of the lifeboat. I would state that when the lifeboat appeared inside of the jetty she was making almost steamboat time, and that against a strong flood tide. This convinced me that they were, and had been, from the time they knew of our plight, doing all in their power for us.*

After getting north side of the wreck—a job that required great courage and skill on the part of Captain Haynes and great strength and endurance on the part of the crew, who, had they not been well drilled, would, I believe, have been exhausted by the long pull they had from the station to the end of the jetty to have performed it—several attempts were made to get us a line before it was accomplished, every attempt at the peril of the lives of the entire crew of the life saving crew.

I must thank you for the lives of all of us that were rescued by the United States life-saving crew, for I believe it was through your efforts that the life saving station was changed from Cape Arago, from where no lifeboat could have reached us that day or the next.[39]

The official inquiry, November 1, 1898, found that Captain Reed had acted in the best manner of seamanship and absolved him of any responsibility for the wreck. Peter Bergman, the wheelman on duty in the pilot house, upheld Reed's report.[40]

Other tragedies on the Coos Bay bar included losses of the barge *Chinook*, April 22, 1907, the *Marconi*, March 23, 1909, the steamer *Osprey*, November 1, 1912,with five lives lost, the schooner *Advent*, February 18, 1910, the steam schooner *Claremont*, May 29, 1915, and the steamer *Santa Clara*, November 9, 1915, with 12 lives lost.[41]

Of all these shipwrecks and losses, none was as dramatic, or as tragic, as the steamer *Czarina*. She wrecked at the mouth of

The U. S. Life-Saving Service moved in 1894 to new quarters on the North Spit of Coos Bay. Crews practiced lifeboat and breeches buoy drills at their isolated post in the sand dunes.

Coos Bay on January 12, 1910. The *Czarina* was a 1,045 ton steam vessel owned by the Southern Pacific Company and valued at $100,000. She headed out of the bay loaded with coal from the Beaver Hill Mine. She also carried 40,000 board feet of lumber on her decks and one passenger. Twenty-three men, including the officers, made up the crew.[42]

The *Czarina* left port at 11:15 A. M. The trip down the long bay was uneventful until she made the last turn in the channel and headed straight for the ocean. Then she began to ship water. The sea was very rough outside and huge breakers slammed at the bar. A number of people thought that a prudent mariner would not have risked crossing such a turbulent bar. Captain W. A. Magee, master of the harbor tug *Astoria*, watched the *Czarina* steam down the bay. His vantage point was a tower in Empire City, four miles from the entrance. He testified that when the *Czarina* had worked her length beyond the black buoy, where the channel turned oceanward, she seemed suddenly to lose headway, stop, and move backward. After a momentary lull in the sea, she went ahead again. Magee said:

> *The ship "seesawed" back and forth for several minutes and then swung her head well to the northward, as if she intended to try for a less difficult passage to starboard. Shortly, however, she swung around to the southwest and went unsteadily forward until she brought up on the South Spit. Then she blew a distress signal.*[43]

Magee counted sixty-one waves breaking over the ship before she struck the South Spit. When the *Czarina* sounded a signal, Captain Magee boarded his tug and started down to the bar to give assistance. By the time he reached the bar, the distressed ship had crossed and was drifting northward. Magee decided not to cross the bar. He explained his actions in the following words:

> *The bar was too rough for us to attempt to cross. After seeing the position of the* Czarina *I knew that nothing could be done from the outside. A steam schooner was off about three-fourths of a mile from the wreck, standing by.*[44]

The steam schooner *Astoria*, loaded heavily with lumber, was the ship to which Captain Magee referred. He saw her move toward the stranded *Czarina* and turn back again, as if the water were too perilous. Later, the captain of the *Astoria*, verified this. Engineer Harry H. Kintzel, the only survivor of the tragic *Czarina*, thought that if the ship had indeed come closer, she might have gotten a line aboard and saved the people. Next morning, while a number were still in the rigging, the *Nann Smith*[45] tried to get near but gave up because of the danger.[46] The *Nann Smith*, a steel steamship owned by C. A. Smith Lumber Company, transported lumber from Coos Bay to San Francisco.

Kintzel said that while the *Czarina* was being buffeted across the bar, heavy seas flooded her engine room and put out her fires. She gained the quieter waters beyond the bar, but by then was entirely helpless. The crew climbed into the rigging while crossing the bar because of the huge waves coming aboard. Once outside the captain dropped the anchor. He thought that he could hold the boat until another ship could get a line to him but the anchor failed to hold.[47]

The ship continued to drift north of the jetty and back into the breakers. Finally, the anchor caught while the ship was still several hundred yards from the beach. At this time the captain, aware of the precarious position, ordered the anchor line cut. The men tried but were forced back into the rigging by the heavy seas. The anchor then held the ship inside several rows of breakers at the mercy of the huge waves. Had the men been able to release the anchor, the ship could have floated closer to the beach and helped the rescue. Held thus, the ship took the full force of the ponderous breakers. Soon, the entire hull sank, leaving only the masts and rigging above the waters. The sea completed the demolition of the deck load, sending it overboard to fill the breakers-swept space between ship and shore. All this wreckage menaced the lives of both the sailors and those trying to save them.[48]

A large crowd from the towns of the harbor soon gathered to try to be of assistance. They built huge bonfires and kept the vigil during the long, stormy night. The crowd stood by helpless and watched victims hanging in the rigging. They saw the men, overcome from the exposure, drop, one by one, and sometimes in groups to their death. They watched as the sea continued to toss the wreckage about and against the masts as if impatient to drive the hapless sailors from their refuge and compete the tragedy.[49]

About dark, the port rigging of the mainmast gave way and swept Engineer Kintzel into the sea. Two or three of his shipmates went with him. He said that he exchanged some words with them after they fell into in the water. He soon became separated from them in the gathering darkness. Kintzel wore a life jacket and thought that some of the others did also. Once the current swept him almost to the beach, but his strength was spent. He could not fight against the strong undertow. The sea carried him near the vessel again. There he got hold of a plank to which he clung even after consciousness left him. His tenacity finally saved him, for the plank dragged him toward the beach, and a surfman waded into the dangerous waters with a line tied to his waist. He tied Kintzel to his line and the life-saving crew dragged them both ashore. Kintzel's ordeal was remarkable in that most victims could not survive the extreme cold for more than an hour.[50]

Many local residents severely criticized Life-Saving Keeper Clarence W. Boice. They thought he had not done all in his power to save the men. Boice saw the *Czarina* as she went down the bay and alerted his crew. Because of the bar conditions, he thought she might meet with disaster. He manned the oar-powered lifeboat and pulled down the channel to the entrance. By then, the crippled ship had crossed the bar and was drifting northward. The keeper then landed near the end of the North Spit on the inside of the bay. Boice gave the following report:

> *I could see that the* Czarina *was just above the north spit. Realizing that I could not reach her with the lifeboat on account of the rough bar, I called my crew out of the boat, enlisted the service of two fishermen who were on their scow boat where we landed, and we all ran to the bar boat house, where I gave the No. 1 surfman orders to get out the beach apparatus cart, and myself went into the bar lookout to observe the action of the vessel. No. 1 got the beach cart out and started along the beach.*
>
> *I noticed that the vessel had anchored just outside the break on the beach. I judged from the trend of her chain*

Men clung to the rigging of the ill-fated Czarina *on January 12, 1910. The vessel stranded in the heavy surf off the North Spit. Only one crewman survived.*

that she had let out lots of cable, and I thought there was a possibility of her holding on, so I phoned from the lookout tower to Mr. Shine at Empire requesting him to wire Superintendent Varney at San Francisco that the Czarina *was anchored just outside the breakers, apparently in trouble, and for him to communicate with the revenue cutter,* Manning *to come to her assistance. I then asked Mr. Shine to notify the company to which the* Czarina *belonged.*

By this time, however, I could see that the ship was drifting to the northward about parallel with the breakers. I remained in the lookout about 20 minutes. I then went down and joined my crew with the beach cart, overtaking them about a quarter of a mile up the beach. We followed along watching the ship slowly dragging, but she shortly brought up, the waves breaking over her bow, then over the body of the vessel, and she swung stern to the beach and came in some. At this time there were three lines of breakers outside the ship and a succession of breakers between the ship and the beach. We stopped with cart abreast of her. I remained with the beach cart and sent No. 1 with some of the crew and 30 or 40 people whom I had asked to assist us, to the bar boat house to bring the surf boat.

They got back in a few minutes with the surf boat. By this time the vessel had foundered and the sea was breaking all over her. I now had the beach cart taken down close to the surf, and just as we were ready to remove the gear the surf came in around the cart and I had to move it back up the beach a little. The Lyle gun was put in as close to the surf as possible and a No. 4 line seven hundred yards long was sent out with a six ounce charge of powder. There was no wind and the line went straight out, but fell 200 yards short. While the crew was hauling in No. 4 line a No. 7 line 600 yards long was sent out with a 6 ounce charge. This line fell considerably shorter than the first. By this time the vessel had settled so much that if a line had fallen across her the crew could not have left the rigging to secure it, as the seas were breaking over her continually and there was no stay between the masts to which the line might catch. Some of the men had already been swept from the rigging. I judged the distance of the vessel to be about 900 yards, and seeing that it was impossible to reach her with the beach apparatus, we manned the surf boat and attempted to launch it. The surf was so heavy however that we could not get off. I was knocked down by a sea while we were trying to get the boat off the wagon, which had been run down into the water. The lumber that had formed the vessel's deck load was coming in by this time so thick that we could not have gone through it even if we had succeeded in getting away from the beach in the boat.

By this time the smokestack had fallen, and the tops of both masts had broken off, and several of the crew had been knocked off the rigging. When we saw that nothing could be done to save the men we scattered along the beach to watch for any who might be washed ashore on lumber or wreckage. About dark the lower part of the main rigging carried away and several of the crew with it. As darkness came on we could see five or six men still clinging to the rigging. Fires were now lighted and the beach was patrolled throughout the night by the station crew, some of whom waded into the surf as far as they dared go with heaving lines tied to their waists. About nine o'clock a man was seen in the surf holding to a piece of wreckage. He was hauled out and found to be breathing. We carried him to a fire and after two hours work restored him so that he could be conveyed to a marine hospital, where he finally recovered.

At daylight of the 13th the mainmast was still standing, and there were still several men in the rigging. The surf still continued as high as on the previous day, and it was impossible to launch a boat. The vessel had settled so that the water was half way up the main mast, and the sea continued to break heavily where she was. The men in the rigging gradually went, and by 10 A. M. they had all disappeared.[51]

At 1:30 P. M. of the 13th, Keeper Johnson of the Coquille River Life-Saving Station, some fifteen to twenty miles south of the scene, reported to Keeper Boice with five surfmen. This crew remained on the beach until the morning of the 14th and assisted in the work of the patrol. Soon they began picking up bodies that floated in to the beach.[52]

The officer who conducted the inquiry of the disaster expressed the opinion in his report that "no blame should be attached to the service, nor to the keeper, nor to any member of the Coos Bay Life-Saving Station, for the deplorable loss of life." In support of this conclusion he cited the obstacles that lay in the way. His statement included the conditions of the storm, the huge seas, and the hazard of the lumber and wreckage that prevented the life-saving crew from launching their surfboat. He also noted that the ship foundered too far from the beach to be reached with a lifeline.[53]

Nevertheless, a number of people in the community pressured the service with questions and accusations for a full investigation. They made eight formal charges against Keeper Boice. No one mentioned irresponsible action of the ship captain during these accusations. Even Boice did not mention it. Boice was exonerated, however. Lieutenant W. W. Joynes, assistant inspector of the seventh life-saving district of Elizabeth City, North Carolina, was assigned to the investigation. At the conclusion Lieutenant Joynes summed up:

> *I wish to direct attention to the testimony of Keeper Boice—His assumption of all responsibility; his bearing throughout the investigation was such as to command the admiration of all who heard his declaration—frank, sincere, and straightforward, and beyond doubt strictly truthful. He is a young man 31 years of age, having been only two years a keeper and in charge of a hard station, with a large, dangerous area under his supervision.*[54]

Public pressure was so great, however, that Boice resigned from the position. He continued in the service but was demoted to No. 1 Man at the Bandon Station where he had served previously.[55]

The Life-Line, *a gospel boat, traveled Coos Bay to carry Rev. Leroy Hall's message to miners, farmers and townsfolk.*

Harold Millis was the one passenger who lost his life along with twenty-three crewmen. Millis, a young man intending to enter a seminary for the Christian ministry, booked passage on the ship to California. His church congregation was among the people who gathered at the beach and watched in horror as the men fell from the rigging into the raging sea. Reverend G. Leroy Hall suffered through the horrifying experience. In watching the futile efforts of the life-saving crew trying to get a line to the ship, Hall became inspired for a different kind of service. He resigned as pastor of the First Baptist Church. He then raised funds and constructed a small boat that he appropriately named the *Life-Line*. Hall contracted with William Cavenaugh to build the craft. For many years he toured the communities of the bay with the gospel message. The first wedding on the boat was July 29, 1913.[56] The boat was 32 feet long and seated as many as 25 people. His unique ministry attracted national attention.[57]

Coos County residents continued to consider San Francisco as "going to the city" over the years. Some travel, of course, was to and from Portland and Seattle. Passenger traffic out of Coos Bay gradually increased to 16,335 in 1908 and by 1914 a total of 21,994 took passage on the sea. With the coming of the railroad from Eugene in April of 1916, travel patterns changed. The gradual decline showed 4,115 in 1916, but by 1927 only nine passengers traveled by boat out of the area. In the early 1920s, rough highways into the area and improved motor vehicles provided a new means of travel.[58]

In 1909 the Oregon State legislature enacted laws to provide for organization of the port. Right away, Coos Bay commercial interests instrumental in passage of the law established the Port of Coos Bay. Voters of the county approved the measure in April, 1909. The first commissioners were Henry Sengstacken, Dr. Everett Mingus, W. C. Harris and W. P. Evans. Without adequate funds they operated under duress of court hearings until 1912. That year, the port went into the hands of a receiver. The citizens reorganized the port with taxing authority. Governor Oswald West appointed Albert H. Powers, Peter Loggie, Henry Sengstacken and Louis J. Simpson as commissioners. The new organization soon sold

COAL EXPORTS FROM THE PORT OF COOS BAY 1860-1915[62]

YEAR	SHORT TONS	YEAR	SHORT TONS	YEAR	SHORT TONS
1860	3,143	1880	46,276	1900	39,602
1861	4,628	1881	31,500	1901	38,303
1862	2,815	1882	30,000	1902	44,842
1863	1,185	1883	31,265	1903	49,906
1864	1,300	1884	35,665	1904	60,150
1865	1,525	1885	29,000	1905	75,785
1866	1,753	1886	51,595	1906	40,003
1867	5,235	1887	35,000	1907	27,192
1868	10,524	1888	35,000	1908	26,629
1869	14,758	1889	44,850	1909	20,000
1870	20,567	1890	69,052	1910	10,055
1871	28,690	1891	40,000	1911	4,746
1872	32,562	1892	36,183	1912	966
1873	38,066	1893	31,245	1913	2,825
1874	44,857	1894	35,665	1914	8,300
1875	32,869	1895	61,277	1915	709
1876	41,286	1896	89,960		
1877	40,000	1897	74,549		
1878	40,000	1898	46,881		
1879	40,000	1899	56,717		

$300,000 in bonds for harbor improvement. Later, commissioners were elected.[59]

In 1987 Senate bill No. 962 made Coos Bay a state port and gave the governor authority to appoint the commissioners. At first the governor appointed seven commissioners with five from the local area and two from outside. Later, the two from outside Coos Bay, from its inception, promoted the industrial base of the area. Such an organization is able to look at all aspects of transportation and how the local economy is affected. The organization also works with the state and federal governments. In the latter years of the twentieth century, the Port made its property on the North Spit available for industrial development.[60]

In 1924, work began on the South Jetty. U. S. Army Engineers estimated the cost would be in excess of $3.5 million. The

The Kronholm family operated the Central Hotel in Marshfield in the late nineteenth century. Workers coming to the mines often booked their first night's lodging at the Central or the adjacent Blanco Hotel. (Kronholm Collection)

By 1890 Marshfield had spread along the upper reaches of Coos Bay. The large bunker of the Newport Mine in the center background created the name Bunker Hill for the community which subsequently developed at the mouth of Coal Bank Slough. (Slattery Collection)

Corps contracted with the Hauser Construction Company to build this breakwater. Construction crews cut a tunnel through Coos Head, through which they hauled rock from the barges over a short line railroad. They completed the jetty November 6, 1928, with the rock causeway extending 4,350 feet into the sea from the Coos Head tunnel. Engineers also repaired parts of the north jetty during those years.[61]

The decline in coal production continued in spite of the significant improvements to the harbor.

COMING OF THE RAILROAD

Ever since the building of the Oregon and California Railroad through the Willamette Valley in 1872, residents of the Oregon south coast dreamed of a rail connection to the valley. A number thought that a larger market would then open for coal in the Willamette Valley and even California. Most of the Portland area coal came from mines in Washington. Promoters in Roseburg and the Willamette Valley also envisioned profits from a railroad to the Oregon Coast. Between 1870 and 1894, a total of sixteen different companies organized in Coos or Douglas counties to enter the possible lucrative transportation business. The Coos Bay, Roseburg, and Eastern organized June 19, 1890, did the most by constructing a railroad from Coos Bay to Myrtle Point by 1893. The project faltered at Myrtle Point for lack of capital; however, the railroad helped the local economy.[63]

In 1886 William and Lloyd Whipple completed a survey of 36 miles from Drain to Scottsburg for the Oregon and California Railroad. Although they found a suitable route down Elk Creek, no action was taken. For several years promoters tried to interest capitalists in this route. Finally in 1905, E. H. Harriman, principal owner of Southern Pacific and thus the old O & C route though Douglas County, decided to extend his railroad to the West Coast. He set up the Oregon Western Railway Company with William Hood, chief engineer of Southern Pacific, in charge of constructing the line 73 miles down the Umpqua to Coos Bay. The crews built

The Coos Bay, Roseburg & Eastern extended east in 1896 from Coos Bay to Coquille to open coal shipments from the Beaver Hill Mine. Ben Chandler, Sr., later a Marshfield banker, sat in front of the lantern atop the first locomotive to travel the route. (Slattery Collection)

bridge abutments and dug two tunnels, neither of which was completed. Later, Oregon highway engineers completed one tunnel and routed Highway 38 through Hancock Mountain east of Elkton. The Oregon Western Railway discontinued the work in 1907 with different rumors for the reason. No doubt, estimates for economic gain influenced Harriman's decision to discontinue the project. Some said, however, Harriman, meant only to thwart the efforts of the Chicago and Northwestern, a company supported by several wealthy lumbermen such as Wilson F. Jewett of Gardiner. Chicago and Northwestern investors were purported to have plans to build a rail line from Pierre, South Dakota, to Gardiner at the mouth of the Umpqua, with headquarters in Drain, Oregon.[64]

Frederick M. DeNeffe, who was involved in planning the railroad line that eventually was built to Coos Bay, penned these words in 1956 about the venture:

The railroad to Coos Bay had its inception in a promotion which, in the beginning, had no substantial backing. At several stages in its development, it was referred to as a "shoe-string railroad," a "shoe-string project," and finally, as the "Mysterious Railroad." Later, it blossomed forth as a railroad project having the financial backing of lords, dukes and other members of the British nobility.[65]

In 1911 the Southern Pacific was forced into action when William J. Wilsey obtained financial backing from a British syndicate headed by Sir Robert Perks. Wilsey was an entrepreneur with more desire than financial backing. His group planned to build a railroad from Portland west to the Pacific Ocean down the Salmon River. From there they intended to go south along the coast to Eureka, California. They also planned a branch line from Eugene to Florence. After months of negotiations following failed attempts with American and French capitalists, Wilsey received "go ahead" orders from Sir Robert Perks. [66]

Wilsey had a letter of introduction from his French acquaintances, but a month passed before he finally received an audience. Perks delayed Wilsey for several more weeks, yet entertained Wilsey in London clubs and Scottish deer stalks. Wilsey enjoyed the entertainment but ran out of funds. Apparently, during this time, Perks was checking on Wilsey and his Paris connections. Finally, Perks permitted Wilsey to show his maps and other pertinent data. Perks liked the idea and took it to his syndicate members. The syndicate was made up of a number of powerful men headed by the Duke of Portland, the richest man in England. The Duke of Norfolk also was a member. The delays continued because a project of that magnitude required a number of syndicate meetings. Weeks went by with little encouragement. Wilsey went broke and could get no more expense money from home. He took a job in a small cafe as busboy. He lived in the back room of the eating place and acted as night watchman. Wilsey continued to meet with the royalty, socially, and at their business meetings. DeNeffe mentioned Wilsey's tenacity and remarked about his situation:

Wilsey was an unusual and interesting character. His four and one-half feet, one hundred pound physique was not im-

pressive; but he possessed suave and charming characteristics, such as a fine profile, a well-proportioned physical make-up, an active and intelligent mind, and refined and courteous manners. He impressed the members of the syndicate with his sincerity and his knowledge of the project and its merits. He also impressed them favorably by his conduct and demeanor at their meetings. His faith in the project and his persistence carried him to success.[67]

The British finally decided to finance the railroad, provided its feasibility was recommended by a competent railroad engineer. The McArthur Brothers of Canada, railroad builders and contractors, recommended H. A. Sumner, an outstanding American railroad construction engineer. Four months later, Sumner went to London with positive approval of the project. Sumner had found out in Portland that James Hill's Electric Railway Company was building a line down the Willamette Valley from Portland and was nearing Eugene. On Sumner's recommendation, the British syndicate decided to defer work on the Portland-Pacific Ocean section and build the Eugene-to-Coos Bay route first. Wilsey and Sumner returned to Portland in the summer of 1911, and there Wilsey deposited $500,000, furnished

The coal-powered locomotive of the Southern Pacific Railroad attracted a crowd on April 5, 1916, in North Bend. (Beckham Collection)

by the syndicate, in a Portland bank. Wilsey used a close friend to write the checks, thus adding to the secrecy of their endeavors. Wilsey bought out the Pacific Great Western Railway Company, an under-financed group of Eugene businessmen hoping to build the line to Coos Bay. The syndicate members were pleased with Wilsey's move. Pacific Great Western had surveyed several miles from Eugene and had acquired rights-of-way through a number of properties. Intrigue followed intrigue with the Harriman interests (Southern Pacific) accusing their old rivals, the Hill family of the Great Northern and Northern Pacific, as being the backers of the "upstart" railway. The Hills denied any connection to the new railroad.[68]

The Southern Pacific acquired the Willamette Pacific Railroad Company which had been formed by the Wendling Johnson Lumber Company. The lumber company held large stands of timber on the Siuslaw. The Southern Pacific took over the surveys and began its fight against Pacific Great Western. There were tense times on the surveying lines and construction sites. Men with rifles protected Wilsey's surveyors. When Wilsey purchased a 40 acre tract of land near the proposed Noti tunnel, the Harrimans knew that they were competing with more than a "shoe string outfit." The British syndicate became a formidable foe. Soon, the Harrimans found out the truth. Sir Robert Perks, while visiting a banker friend in New York, was so proud of his project that he drank a little too much and spilled the news to the banker. The banker was a close friend and ally of the Harrimans. The race continued but Pacific Great Western met the legal requirements ahead of Willamette Pacific. The law was quite specific in its terms. The company with the first completed and accepted survey had priority to the area.[69]

Engineer Ralph J. Hunt, working for Pacific Great Western, soon found himself competing with his own survey. Earlier, the Pacific Great Western, had tried to interest the Southern Pacific in the line and had presented that company with Hunt's survey data west from Eugene. The Southern Pacific, through its subsidiary, Willamette Pacific, rushed headlong into the fray. They crisscrossed Hunt's work and attempted to get ahead of the Pacific Great Western. Both companies knew the rules for prior rights. Just at dark one stormy night, Hunt completed his work and mounted his horse for Eugene. He hurried for twenty miles with his data through the storm and mud. Hunt arrived in Eugene at three o'clock in the morning. His board of directors held a special meeting and accepted the survey. His rivals had to go to Portland. Pacific Great Western won by a few short hours. Although it took some court proceedings over the next several weeks, Pacific Great Western's prior certification proved them the winner.[70]

After a number of suits and counter suits, the Harrimans gave up the fight with a purchase offer. Apparently the British interests felt the offer was too good to refuse so they accepted. They were reluctant to sell, however. Perks told Wilsey, "the project had just gotten to the state where the fighting was beginning to get good and interesting." The valiant and redoubtable Wilsey, in his determination to succeed, forced the issue by demanding that the sale contract include the completion of the line to Coos Bay in a restricted amount of time. On July 1, 1912, William J. Wilsey was guest of honor at a banquet in North Bend and lauded as a hero for his efforts.[71] It took four more years to complete the long-awaited railroad.

Hopes for prosperity leaped with the arrival of the Southern Pacific Railroad on April 5, 1916. The Marshfield City Council gave a right-of-way down Front Street where a throng greeted the first train. (Stadden Collection)

In August of 1915, the Marshfield Chamber of Commerce sent letters to a number of towns in the valley promoting the sale of coal. The railroad had opened an important new market for southwestern Oregon coal. The letter stated, "When the Willamette-Pacific is completed between Coos Bay and Eugene, Coos County can furnish some of the best of coal for domestic uses at very competitive prices."[72]

Periodic reports on the progress were printed in the local papers. The *Coos Bay News* of December 7, 1915, reported:

> *At present the line is being constructed to complete the section between Eugene and Cushman, the latter point being in the vicinity of Acme. At this point the railroad makes it seventh and last crossing of the Siuslaw River, proceeding thence south to Marshfield, paralleling a chain of lakes between the Siuslaw and the Umpqua Rivers and the north inlet of Coos Bay. Grading of the line is entirely completed including the tunnels, of which there are nine. At present the construction is concentrating on the eighteen miles between tracks from the bay to meet that from Eugene.*[73]

Not until April of 1916, however, did traffic commence on the long-awaited railroad. For the next seven decades the railroad experienced increased use. With the closing of sawmills in Coos County in the 1970s and 1980s, rail hauling out of the area declined. Highway trucking and fewer customers gave the Southern Pacific serious financial concerns. Government regulations and choking union contracts added to the problems. Other factors included the condition of several bridges. The costly replacement of the long span across Coos Bay was the most serious.[74]

In November of 1994, Southern Pacific entered into a lease purchase contract with Central Oregon & Pacific Railroad Inc., a subsidiary of RailTex, Inc., to take over the Eugene to Coquille line along with a portion of the Siskiyou line that extended 336 miles from Eugene to just south of Ashland. RailTex, a Houston, Texas based firm, budgeted $600,000 for repairs to the Coos Bay Bridge.[75]

The improvements of the harbor and shipping facilities enhanced the profits for exporters of coal. The railroad was a boon to the area also; however, by 1916, the coal market had so eroded

"Drummers," or salesmen, peddled their wares to Coos Bay residents early in the twentieth century. To travel east to Roseburg they took the handcar from the head of Catching Inlet to the stagehouse at Sumner. (Stadden Collection)

on the West Coast that very few southwestern Oregon colliers shipped coal. The Southern Pacific, by that time, owned and operated the Beaver Hill mines so it shipped coal on the new railroad. For a few years, the Alpine Coal Company of Riverton also shipped coal on the Southern Pacific to Eugene and Portland.[76]

Changes in transportation—the lighthouses, the life-saving service, jetties, dredging, harbor range lights, and the railroad—enhanced the potentials for profits. The real challenge lay in the viability of markets for the millions of tons on soft coal in the Coos Bay and Riverton coal fields.

ENDNOTES

1. *Coos Bay News*, August 4, 1915.
2. Ibid., August 14, 1915.
3. Ibid., December 14, 1915.
4. Stephen Dow Beckham, *The Simpson of Shore Acres*, p. 14.
5. *Coos Bay News*, August 3, 1915.
6. Ernest L. Osborne, *Wooden Ships and Master Craftsmen*, p. 23.
7. *Coos Bay News*, August 4, 1915.
8. Ernest L. Osborne, op. cit., p. 2.
9. Alice H. Wooldridge, *Pioneers and Incidents of the Upper Coquille Valley*, p 274. Giles wrote, "We finally got out boat rigged and ready for sea about May 15, 1854, two white men, myself and pet Indian, being the first white men that ever crossed the Coquille Bar.", p. 279.
10. Ernest L. Osborne and Victor West, *Men of Action: A History of the U. S. Life-Saving Service on the Pacific Coast*, p. 39.
11. *Coos Bay News*, August 4, 1915.
12. Stephen Dow Beckham and Rick Minor, Cultural Resource Overview of the Coos Bay BLM District, p. 225.
13. Ibid., p. 228.
14. Ernest L. Osborne and Victor West, op. cit., p. 5.
15. Stephen Dow Beckham and Rick Minor, op. cit., p. 226.
16. George Baxter Case, The History of the Port of Coos Bay, 1852-1952, p. 19.
17. Ibid., p. 23.
18. Ibid., pp. 24-33.
19. Ibid., p. 36.
20. Ibid., p. 4l.
21. Stephen Dow Beckham and Rick Minor, op. cit., p. 228.
22. George Baxter Case, op. cit., p. 49.
23. Ibid., pp. 43, 56.
24. Stephen Dow Beckham and Rick Minor, op. cit., p. 227.
25. Ibid.
26. *Coos Bay Harbor*, June 8, 1907.
27. *Coos Bay Harbor*, February 23, and July 17, 1917.
28. *Coos Bay News*, December 14, 1915.
29. *Annual Report of the United States Life-Saving Service*, 1897, p. 37.
30. Ibid.
31. Ibid.
32. Ibid.
33. Ibid., p 38.
34. George Baxter Case, op. cit., Illustration No. xvii shows the government dock at the tip of North Spit across from Fossil Point.
35. *Annual Report of the U. S. Life-Saving Service*, 1897, op. cit., p. 38.

36. Ibid., p. 39.
37. Ibid.
38. Ibid.
39. Ibid.
40. *Coast Mail*, November 5, 1898.
41. *Coos Bay News*, December 14, 1915.
42. Ernest L. Osborne and Victor West, op. cit., p. 128.
43. *Annual Report of the U. S. Life-Savings Service*, 1910, p. 60.
44. Ibid., p. 60.
45. The *Nann Smith*, owned by C. A. Smith Lumber Company, carried lumber to San Francisco Bay.
46. *Annual Report of the U. S. Life-Saving Service*, 1910, op. cit., pp. 55-60.
47. Ibid., p. 60.
48. Ibid., p. 61.
49. Ibid.
50. Ibid.
51. Ibid., p. 62.
52. Ibid.
53. Ibid., p. 63.
54. Ibid., p. 65.
55. Ibid.
56. Victor C. West, *Southern Oregon Mosquito Fleet*, 1853-1948, p. 26.
57. Anna M. Beckham and Dr. Everett W. Curry, *The First Baptist Church, 1879-1979*, p. 21.
58. George Baxter Case, op. cit., p. 108.
59. Ibid., p. 55.
60. Martin L. Callery, Interviews with Dow Beckham, March 16, 1993, and November 7, 1994) (The Corps of Engineers maintained a depth of 26 feet on the bar and in the harbor beginning in 1945. In 1977-1978 the Corps deepened the harbor entrance to 45 feet at mean low water with a 700 foot wide channel. Beginning at channel mile 2 the Corps began maintaining a channel depth of 35 feet for the next 15 miles. They also maintained turning basins at mile 8, mile12, and mile 14.6. This program was still in effect in 1994.)
61. Charlotte Mahaffy, *Coos River Echoes*, p. 26.
62. George Baxter Case, op. cit., p. 104.
63. Stephen Dow Beckham, *Land of the Umpqua*, p. 202.
64. Ibid., pp. 202-203.
65. Fredrick M. DeNeffe, *The Mysterious Shoe-string Railroad*, p. 210.
66. Ibid., p. 208.
67. Ibid., pp. 207-208.
68. Ibid., p. 209.
69. Ibid., p. 210.
70. Ibid., p. 217.

71. Ibid., pp. 226-227.
72. *Coos Bay News*, May 25, 1915.
73. Ibid., August 15, 1915.
74. Allan E. Rumbaugh, General Manager of the Port of Coos Bay, Speaking before the Riverman's Club, January 25, 1993.
75. *The Oregonian*, Portland, Oregon, November 25, 1994.
76. Karl Alpine, Interview with Dow Beckham, April 15, 1992.

FOUR

MINES, PIONEERS, & EARLY DAYS

Investors operated a total of six coal mines successfully in the Newport Basin. By the 1880s the Newport Mine produced an excess of 30,000 tons a year. The Eastport Mine shipped a like amount to the California markets. The Englewood, South Marshfield, and the Flanagan and Reservoir mines supplied coal for local use. James Flanagan operated the Flanagan and Reservoir mines exclusively for the Marshfield Water Company.[1]

The Englewood Mine. The Englewood and South Marshfield operated within the SW1/4 SW1/4, section 34, T25S, R13W. They were just west and south of Marshfield (Coos Bay) at an elevation from ninety to two hundred feet. The Reynolds Development Company operated the Englewood Mine from 1927 to 1932 and produced about forty tons a day. A. Y. Meyers managed the mine during those years. In 1913 Meyers came to Coos Bay from Spokane, Washington. During the Depression, Meyers extended credit to a number of employees. They did not forget his kindness and consideration toward their financial problems. Years afterward, many of them brought him produce from their gardens in appreciation.[2] The mine produced a total of about 50,000 tons. The record showed seventy acres mined at sixty percent recovery. The Oregon State Department of Geology and Mineral Industries estimated that Coos Bay mines recovered approximately sixty percent of the available coal.[3] Thurston

"Sticker"Jacobson's father hauled coal from the Eastport Mine and sold it to local households.[4]

The South Marshfield Mine. The South Marshfield Mine portal was within the NW1/4 SW1/4, section 34, T25S, R13W at an elevation of about 200 feet. The adit was about a half-mile west of the end of Elrod Avenue. Blossom Gulch bounded the property on the north and the beds reached south to the Englewood workings. The South Marshfield Mine opened in the early 1880s. In 1909 geologist J. S. Diller reported the operators had taken an acre of coal from the eight foot bed of the mine. Allen and Baldwin, in their 1943-44 survey, estimated that the miners had taken coal from a little less than two acres. Tests showed that the vein contained layers of shale, clay, bone, coal, and more shale in a bed eight feet four and one-half inches thick. The coal contained 14.8% moisture, yet produced 8,700 B.T.U. heat intensity.[5]

All the coals in the Coos Bay Quadrangle are sub-bituminous or lignitic. Sub-bituminous coal weighs about 80 pounds to the cubic foot or 25 cubic feet to the ton. Chemists base the classification of sub-bituminous and lignitic coals on their heating value and their weathering and agglomerating properties. Coals of these ranks weather or slack on exposure to the air after mining. They do not form a coherent button in the analysis process. Furthermore, they have a heating value of less than 13,000 B.T.U. on the moist mineral-matter-free basis. The moist mineral-matter-free B.T.U. of sub-bituminous "A" coal is 13,000 to 11,000. Subbituminus "B" coal runs 11,000 to 9,500, and lignites are less than 8,300.[6] In 1911 geologists J. S. Diller and M. A. Pishel reported a bed of coal near the Marshfield Mine. This bed was 50 feet, stratigraphically, below the Newport coal bed.[7]

The Resevoir Mine. Geologists noted outcroppings west of the end of Central Avenue in Marshfield (Coos Bay). In 1889 James H. Flanagan, son of Patrick Flanagan, opened the Reservoir Mine to supply coal for the city water pumping plant. The water company bought coal from Flanagan about fourteen years. Two portals of the mine were just east of the center of S1/2 SW1/4, section 27, T25S, R13W. They opened at an elevation of 125 feet in the north fork of Blossom Gulch.[8]

Earlier, C. A. Merchant opened two tunnels and operated the mine a short time in this same vicinity. The southern tunnel was about one hundred feet long, and the northern tunnel 300 feet long. Merchant mined the block between the two tunnels and produced about 400 tons of coal in the two years that he operated the mine.[9] In 1902, when the coal "pinched out," the workers moved to the Flanagan Mine on the opposite of the ridge.[10]

The Flanagan Mine. The water company used the Flanagan Mine about ten years, from 1902 to 1911, to provide energy for the city water pumping plant. Flanagan opened the portal 600 feet east of the west quarter-corner of section 27, T25S, R13W. When the water company discontinued mining, they leased the mine to a Mr. Swanson. Swanson continued to mine up to the dip to the west, taking coal remaining between the old workings and the outcrop. The mine produced between 20,000 and 25,000 tons of coal.[11]

The Wilcox Mine. In 1911 geologists J. S. Diller and M. A. Pischel examined a mine in the bluff just west of the Weyerhaeuser chip facility in North Bend. The tunnel ran west under the Mercy Hospital. Miners opened the adit 50 feet west of Highway 101 in section 15, T25S, R13W. The bed dipped 35 degrees northwest. Diller and Pischel found the upper coal sixteen inches thick. Twenty-two feet below, they checked a twenty-one inch vein. The miners removed coal from both veins out of the same tunnel. Diller and Pischel found the coal comparable to the other coals in the basin. Local households used the coal during the short time the mine operated. The next outcropping north of the Wilcox Mine was on Simpson Heights. Freeman Lockhart mined the first coal there in 1853.[12] The Newport Basin also extended west to include the Empire and the Oldland Mines. The Empire Mine was near the southwest corner of section 20. The Oldland Mine was on the east side of Joe Ney Slough in the center of the W1/2 SW1/4, section 8, T26S, R13W.[13]

Miners wearing head lanterns stood at the entrance of the shaft house to the Wilcox Mine. The coal deposits extended under the city of North Bend from the portal west of Highway 101 near the Weyerhaeuser Timber Company plant. (Beckham Collection)

Black Diamond Mine. Miners drove a tunnel thirty feet in coal on the point between North Slough and Haynes Inlet in the northwest corner of section 34, T24S, R13W. The coal sloped northeast 17 degrees and dipped 29 degrees west. The bed contained eight feet of poor quality coal. Several other outcroppings of coal show on the north side of the bay. A four foot thick vein surfaced at Lone Rock, in the center of SW1/4, section 25, T24S, R13W.[14] Two veins broke through in the ravine east of the home of George and Gena Stovall along the road to Haynes Inlet.[15]

Glasgow (Hardy) Mine. James Flanagan, son of Patrick Flanagan, was quite successful in coal mining. Flanagan commented on investors and would-be coal miners:

> *Based on many coal prospects and out croppings, promoters would organize companies, sell stock or induce men to 'jump into the venture,' without adequate investigation. Merely upon the theory that vast amounts of good coal lay buried just beneath the surface, waiting to be mined and taken to market, reckless expenditures would be made for surface developments of various kinds. Then followed the opening of the mine only to find in too many instances and too late that the vein was broken, or the coal was of inferior grade, or mining conditions were so unfavorable that successful operations could not be carried on.*[16]

In the 1870s Hardy and other speculators opened a mine on the north side of Coos Bay in the NE1/4, SW1/4, section 1, T25S, R13W. According maps by L. Wagoner (1889) and C. F. Besse (1932), miners tunneled 700 feet north along the bed. At that point, they opened an air shaft to the surface. The coal struck nearly north and dipped 20 degrees west. They made the tunnel large enough for a double track. The seven foot bed "pinched" out before reaching through the mountain to Willanch Slough. In 1911 geologists Diller and Pishel indicated that the coal was inferior and layers of impurities made mining too costly for any expected financial success. Hardy and his investors lost over $100,000 in this venture.[17] Mary Jane Alford purchased one hundred shares of stock in this venture on August 2, 1872, at $100 per share.[18] Alford received no return on her investment. Her only solace was a handsome stock certificate.

Victor West, a North Bend historian, gave the following account:

> *Henry Haynes discovered an outcropping of coal on the east side of Jordans Point, across Coos Bay from old North Bend. Jordans Point and Jordans Slough were later renamed Kentuck Point and Kentuck Slough. Haynes spent several hundred dollars trying to develop his discovery into a coal mine but finally gave up. Judge D. L. Watson*

acquired the land and soon sold it to Hardy and Goodale for $14,000.00.

Hardy and Goodale dug a tunnel into the coal and built other facilities in 1871. The shipping wharf ran out from the southern end of Jordans Point to the channel of Kentuck Slough. The tunnel was about eight hundred feet long and when they struck coal they established a gangway off the tunnel four hundred feet long. The coal was very soft and air slacked badly. They made three shipments to San Francisco. The first shipment was on the schooner Gotoma. *She was loaded with two hundred eight tons of coal and sixty-five thousand board feet of lumber for Asa M. Simpson. After expending $100,000, Hardy and Goodale gave up in 1873. Other entrepreneurs attempted to open the mine at 180 feet below Hardy's tunnel. They took out 150 tons and gave up. Patrick Flanagan acquired the property in 1889 but never attempted to mine it.*[19]

The Gilbertson Mine. In 1891 miners opened the Gilbertson mine on Kentuck Slough. This 10 foot bed of coal in five layers also failed. Miners had some success between 1908 and 1913. D. L. Buckingham surveyed this mine in the early twentieth century. He located the tunnel in the NW1/4, section 6, T25S, R12W. Buckingham observed coal outcroppings in the

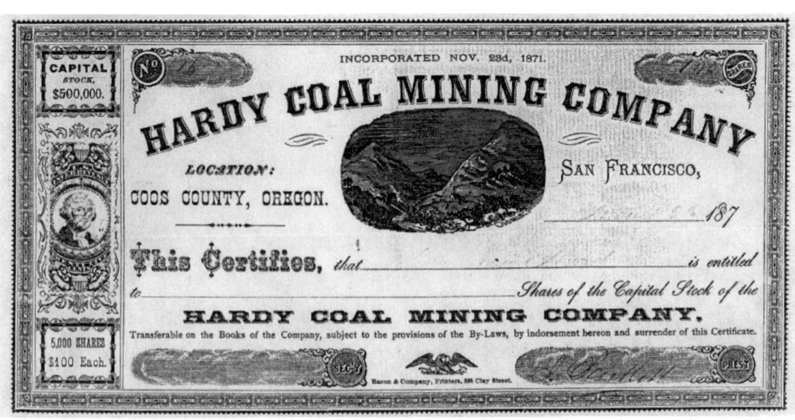

Certificate dated August 2, 1872, for 100 shares of stock purchased by Mary Jane Alvord in the Hardy Coal Mining Company. J. Pashell, president, signed the certificate. (Chaney Collection)

successive creek beds at 127, 227, 236, and 217 feet from the Kentuck Slough road on the south of the township corner. The miners kept no records of the total tonnage taken from this mine.[20] The owners established a bunker on the bay near the mouth of Kentuck Slough and transported the coal from the mine on a small railroad.[21]

Geologists and engineers also recorded Carlson coals on the south side of Kentuck Slough. South of Kentuck Slough, geologists recorded Willanch Slough coals, Ward prospect, Wirth prospect, Smith Mine and a number of other outcroppings.[22]

The Lillian Mine. From 1900 to 1905 Louis J. Simpson operated the Lillian Mine near the confluence of Catching Slough and Coos River. Lillian Slough is approximately one mile up Coos River from Catching Slough. Simpson opened the mine primarily to provide coal for his locomotives. His train hauled logs out of the Daniels Creek area for his sawmill in North Bend. The Lillian Mine opened 700 feet south of section line between sections 4 and 33, T25S and 26S, R12W. The Black Diamond Mine in the same area, was about 800 feet down the creek from the opening of the Lillian Mine. These two mines had five foot beds of coal interspersed with four inches layers of clay.[23] Simpson took out about 4,000 tons of coal from his Lillian Mine. He hauled coal on a narrow guage railway about a mile to Coos River. Workers loaded the coal on a barge and towed it to the railhead at the mouth of Daniels Creek.[24]

Catching Inlet Coals. Coals along Catching Inlet included Gunnell, Messerle, Smith-Powers, Huntley, and West Prospect outcroppings. Other veins showed near the head of the inlet, namely Standley, Noble Creek, Sumner, and Newcastle Mine.[25]

There are two beds of coal on Noble Creek in SE1/4 SE1/4, section 36, T26S, R13W. In 1909, Smith-Powers Company, logging in the area, mined the easternmost bed on a limited basis for its locomotive.[26]

Two local miners, J. B. Sneddon and Charles Sneddon, had a narrow escape in October of 1909. Smith-Powers Logging Company engaged them to open the mine. About noon they discovered the first traces of gas in the mine. They burned out the gas and figured that no problem existed. Generally, little gas formed above sea level. They used chutes to let down the coal. J. B. Sneddon remained near the chutes with the miners' lamps while Charles went ahead to test the gas. In a few minutes, the gas caught and Charles rushed back in a mass of flames. Jack rushed to his brother's aid, but his clothing also caught fire. Then, in another instant, the explosion occurred. They barely escaped death, first from the explosion of the methane gas, and later from the deadly gas, after-damp. After-damp developed after the methane gas exploded. The men were quite badly burned from the explosion and Jack Sneddon suffered a scalp wound. Fortunately, they were not injured seriously from the blast but in the process of putting out the flames and struggling to the opening, they breathed too much of the after-damp. The after-damp eliminated the oxygen so the men were almost overcome. Loggers took them to Marshfield where Dr. George Dix looked to their injuries. [27] Geologist Pishel indicated that Smith-Powers only mined about two acres of coal from the three-foot vein.[28]

Bunker Hill Mine. The mine on the north side of Bunker Hill was at an elevation of about 200 feet, about a thousand feet south of the highway in the SE1/4, section 35, T25S, R13W. Only the top bench was uncovered so the 1943-44 study reported only that the coal bed sloped north 37 degrees east and dipped 14 degrees southeast. This mine, like many others around the bay, may have been opened only for family use or local consumption. Yet it may have been inspired by possibilities of finding a rich vein large enough to attract capital for full scale development.[29]

Andrew J. Davis Mine. In 1856 Andrew J. Davis opened a mine in the ravine about 1,000 feet east of the Isthmus Slough bridge. Davis built a storehouse, a wharf, and a railroad from the mine. After spending $75,000, Davis found the vein unreliable. Davis also had other interests in the area. Davis and James C. Tolman

Lorenzo Dow Kinney (left) and an assistant marked a large coal vein of the Hardy or Steva Bed about 1905 on Kentuck Slough. Kinney platted Bangor at the mouth of Pony Slough and promoted the town as the terminus of a railroad he hoped to build to Salt Lake City. (Slattery Collection)

claimed the site of downtown Marshfield under the Donation Land Act. Tolman sold out to Davis and left the area. Davis hired Wilkins Warwick to manage his properties. Davis and Warwick submitted claim for the property and Davis paid a $1.25 an acre in lieu of the five years occupancy rule. The land office delayed the claim. Davis and Warwick dated their claim August 4, 1854. Congress had changed the law, July 17, 1854. The law disallowed a town site under the Donation Act. Davis sold his interest to Henry Heaton Luse and moved to Iowa and later to New York.[30] Luse paid Davis $12,500 for his interest in the claim. By 1877, when the town had grown to 300, inhabitants were concerned about title to their property. They knew the title was under a cloud and challenged Luse. Luse hired George H. Williams, Senator from Oregon, to represent him. Williams obtained a decision in Luse's favor. There was considerable tension. The people had spent money on their homes and Luse threatened to take over the property. Fortunately for them, Luse sold all of his Coos Bay area holdings and moved to San Francisco. The purchaser, Southern Oregon Improvement Company, made satisfactory arrangements and the people secured titles to their homes.[31]

A number of years later, "Grandpa" Sam Archer worked the Davis mine and gave it his name. Archer also found the mine financially unsuitable. Henry Sengstacken opened a mine in that same ravine. His endeavor also failed.[32]

MINING FAMILIES' HISTORIES

There are many families in the Coos Bay area who can boast a proud background of pioneer spirit similar to the Archer family and their relatives. Following the arrival of the Coos Bay Commercial Company and Patrick Flanagan, many pioneers came because of the lumber mills or homesteads. The Archers, Monroes, and Brynns immigrated specifically because of the coal industry. Their story is representative of many who immigrated to the area.[33]

The Monroes. Neil Monroe (Monro) was born in Scotland July 7, 1818, in Cathcart, Renfrew County. Monroe was one of several children born to Alexander and Euphemia McCallan Monro. Renfrew County in Scotland was in the central lowlands and contained considerable mineral wealth and vast coal deposits. Coal mining was the major industry and the Monroes were a coal mining family.[34]

Neil Monroe, like his brothers and sisters, worked in the mines; however, as a young man he began studying for the ministry. Monroe met and fell in love with the daughter of an aristocrat of the landed gentry. Her family did not approve of her marrying a poor minister, especially one they considered inferior, both in class and wealth. Without their permission, marriage was impossible. So Neil left his love to pursue his fortune in America. He immigrated to Kentucky where two of his uncles had settled. The uncles farmed cotton and tobacco, and they also mined coal.[35]

Neil Monroe acquired land, a couple of slaves, and laboriously began construction of his home with the dream of marrying his lady in Scotland. He diligently worked along with his slaves and gained their respect and devotion. In a few years, he owned a large group of slaves and a beautiful mansion. He built his home in the Ohio River valley, encircled by rolling hills of oak, black walnut, hickory and other hardwoods. The home had many rooms with a two-story entry paneled with black walnut.[36]

In 1842 Monroe returned to Scotland. He soon convinced his sweetheart's parents that he was wealthy. That made him acceptable so they got married. Her parents bought them furniture for a wedding gift, enough to furnish their entire home. They had three daughters, Euphemia "Lucy," born in 1844, Elnora "Ellie" born in 1846, and a daughter who trained to sing in the opera but died of pneumonia before she was twenty-years-old.[37]

The mother and daughters made several trips home to Scotland to visit the grandparents. On one trip they engaged a governess to teach the girls. Elizabeth, the governess, returned to America and enjoyed a loving relationship with the girls. The mother had always been sickly and died when the girls were still young children. In 1858 Neil married the governess of his daughters. This union produced two children, Jeanette, born in 1859, and Margaret, born in 1865.[38]

Because of the great war between the states and the failing cotton market, Monroe lost most of his wealth. Life in Kentucky, after the Civil War, was very unsettled. Monroe received word from a cousin in Oregon, who painted glowing pictures of vast resources and opportunities in the Pacific Northwest. Monroe decided to move his family to the West. In 1871 he and a partner bought a ship, a fore-and-aft-rigged boat with one mast and a single headsail jib, known as a sloop. They floated down the Ohio to the Mississippi River and on to the Gulf of Mexico. From there they sailed around Cape Horn and north to Oregon. After six months, they sailed through the Coos Bay harbor entrance and soon sighted the first settlement of the Coos Bay area.[39]

In the early 1870s Empire City possessed a crisp newness with its white clapboard businesses clustered along the waterfront. Virgin forests of Douglas Fir and spruce encircled the town.

The contrast between this frontier and the Ohio River home of the Monroes was immense. Their fears of the unknown, however, were overcome by the promise of hope in these unsettled lands. Several communities already existed in the Coos Bay area. Marshfield was developing as a lumbering and shipbuilding town. It was built on piling over the wetlands. Mill Slough ran through the middle of the community. Small boats and barges navigated Coal Bank Slough loading out coal for the San Francisco markets.[40]

Asa Simpson operated his sawmill and shipbuilding yards near the bend of the bay where his son, Louis, later platted the town of North Bend. Many farmers settled the Coos River, Catching Inlet, and Sumner areas. Bay City and Eastside were other communities. Coal mining was a major industry. The Newport and Eastport mines were in their heyday. Other mines opened from time to time. Some prospered and some failed. Later the Southport Mine, Delmar Mines and Beaver Hill Mine began operations.[41]

With previous experience in the mines, Neil Monroe soon started working at the Newport Mine. He took the sloop up the Coal Bank Slough to Libby and unloaded their belongings for

Coos Bay families sometimes found quiet moments in summer to relax from their labors.

their new home. Libby grew up haphazardly along the tracks that led from the Newport Mine to Coal Bank Slough. Picket fences protected the children from the locomotive pulling lines of cars through the village. Monroe built a small house just east of Libby. He papered some of the walls his worthless Confederate money.[42]

Monroe brought grist stones from Kentucky. The area farmers had sent their grain to San Francisco to be milled, so the grist mill was a welcome addition to the community. In 1860 Henry Schroeder built a grist mill at the Baltimore Colony in the Coquille Valley, near what later became Myrtle Point. W. T. Lehnherr also constructed a grist mill in the Coquille Valley.[43] It was easier for Coos Bay area residents to have grain milled in San Francisco than in the Coquille Valley because of the difficulty of traveling to and from the valley. Ships went regularly back and forth to California. The Monroe family settled into life on the Oregon Coast and became active in the community.[44]

The various communities remained somewhat isolated, since travel was primarily by boat. Each community had its 4th of July celebration, Christmas tree, an annual community Christmas Eve party, and pie socials. Picnics and Saturday night dances

Men and boys found good fishing for trout and salmon on the Coos River and other creeks pouring into Coos Bay a century ago.

were common. Each community boasted a brass band and a baseball team. Social activities were plentiful. Monroe and his wife, Elizabeth, hosted parties on their sloop. They brought large groups up from Marshfield to picnic, swim, and visit for hours. However, one day Monroe went to the wharf in Marshfield where he had docked the sloop. To his dismay, he discovered that his partner had stolen the ship and left, never to be heard from again.[45]

In the years prior to 1916, Oregon's valley towns, and even Portland were more remote from the Coos Bay area than San Francisco, in part because of the constant shipping to San Francisco. Ships left weekly for the "city" and passenger fare was $10. *The Coast Mail*, a local newspaper, reported the Monroe family trip on the steamer, *Success*. Many other families considered going "outside," the zenith of vacations.[46]

A few years later, the Monroes moved to South Marshfield near the intersection of Third and Johnson streets. Elizabeth and her daughters were active in the community. Elizabeth belonged to the Literary Society, a cultural club that pursued a variety of programs such as debates on issues of the day, lectures, musicals, and drama. She worked on committees for the July 4th celebrations and the annual Christmas tree held at the Academy. The Academy was a private high school in Marshfield. Elizabeth was very much the lady. She may have lived in a lumber and coal mining town, but she expected her girls to be educated and proper. Ellen remembered Elizabeth telling her, "You may live on a farm but you're not a plowhorse," and made her walk around the room with a book on her head, practicing the posture of a lady.[47]

Neil Monroe was a quiet, serious man who was deeply religious. He was kind, gentle, and always fair to those around him. He never became a minister but remained religious all his life. Ellen remembered his long prayers and how he presided at services for a large following. He led them to the Coal Bank Slough where he baptized them. They took off their shoes and waded into the water for the baptism. Between pranks, Samuel Archer courted Monroe's daughter, Margaret. For a prank and some his fun, Archer spread the rumor that he had placed some traps in

 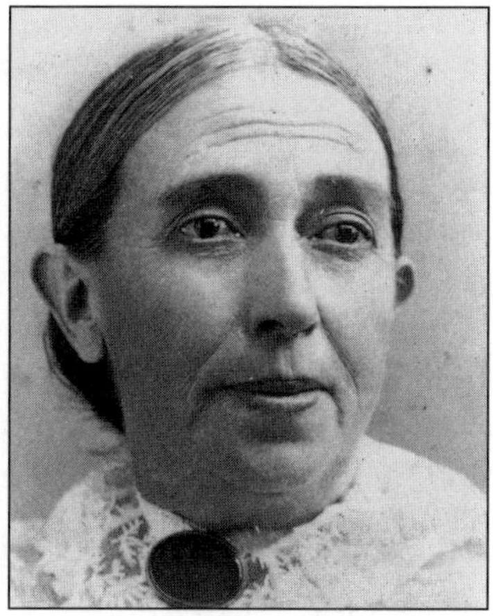

Neil Monroe (1812-1902) and his wife, Elizabeth Monroe (1828-1904), parents of Margaret Monroe Archer.

the slough. Monroe heard of it and searched carefully but did not find any traps. It took him a while to forgive Sam.[48]

All the Monroe daughters grew up in Marshfield and married local men. Ellie married Frank Reiby and they had three children. Lucy married a Butler and moved to Tacoma, Washington. Jeanette and her husband, Charles Bonebrake, whom she married in 1885, had six children: Clara, Zelle, Malissa, Lucy, Charles, and Holt. Margaret married Samuel Archer and helped him homestead in Bay City. Bay City was the name applied to the area south along Isthmus Slough near the present bridge to Eastside. In their later years, Neil and Elizabeth Monroe lived near Margaret and Sam Archer on their ranch in Bay City, a farm known as The Red Barn.[49]

The Archers. Samuel Archer was the second child born to William and Martha (Littlewood) Archer of Lanark County, Scotland. He was born in Newton Hamilton, Armaugh County, Ireland, in 1843, while his parents were there visiting relatives. He reminded his children often of this fact but took pride in being a Scotsman. His older sister, Rachael, was born in 1841 in Scotland. Their mother died when young Sam was only five years old and William was left to cope with a trying to raise a very mischievous son.[50]

Young children worked in the mines in Scotland for many years and Sam did not escape this early introduction to mining. William started Sam when he was only six-years-old. Many coal beds in Scotland were only eighteen to twenty-four inches deep, and men had to lie down to dig coal. They used young boys to help in the mine. It was very hard work, but Sam developed a robust physique and a strong and resolute character. His stern father often punished him, but perhaps justifiably, because Sam was rebellious.[51]

Sam Archer began pulling outrageous escapades at an early age. For example, he sneaked on to the large estates and caught fish from the stocked ponds. He hid the fish in his baggy pants and sold them in the village. Poaching was a serious offense, but Sam was too sly for the authorities for he never got caught. William found it more and more difficult to cope with the rebellious boy, so he sent Sam away to live with his sister who kept her nephew for a few years.[52]

Sam Archer did not see his father again for a long time, nor did he live long with his aunt. He signed as a cabin boy on a ship leaving Scotland and planned to "jump ship" when he reached America. He had heard of this "great land of opportunity" and wanted to leave Scotland and the bitter experiences of his youth. He made several attempts to leave the ship, but each time the captain caught him and took him back. On one trip in his early teens, he thought he was away free. Sam and a young friend left the ship and hid out for several days in Montreal. Finally, cold and hungry, they came out of hiding. As they turned a corner they met the captain face to face. Learning to fend for himself early in life and coping with stern ship captains made Sam hard and tough for his age. His resiliance was only equaled by his bellicose precocity. Undaunted, and without a moment's hesitation, Sam reached up and pulled the captain's stovepipe hat down over his eyes and ran. Both boys ran, still determined to leave the ship. The captain yelled for help and soon he took the boys back aboard where they were confined until the ship was well out to sea. During those years Archer sailed to many ports around the world, but he continued to dream of getting to stay in America.[53]

In the fall of 1856, when Archer was about sixteen, he and his friend again signed on as cabin boys. This time they formulated a sure-fire plan of escape, or so they thought. Their ship sailed up the St. Lawrence River to the port of Quebec. The lads were ready. When darkness came, they threw a rope over the side of the ship and slid down into the freezing water. They intended to hold onto the rope until it was safe to swim ashore. They had not known how quickly the cold waters would affect them. They were not able to climb back up to the ship and were very frightened of being caught. Archer told his children of this night many times. It haunted him all his life. The last words he heard from his friend were in despair, crying out "Sammy, I can't hold on any longer." Sam himself also drifted into unconsciousness. He did not know how it happened, but he awakened later in an Indian camp on the shore. Sam spoke often of their kindness and patience. Sam carried a knife in his belt, so he gave the knife to a young Indian in appreciation for having saved him. The Indians were attracted to Sam's red hair and wanted him to stay and live with them. In a few days, however, Sam decided to move on and pursue his dreams. [54]

Archer traveled to Collingwood, Ontario, visiting uncles and cousins, never staying long, still looking for adventures. He found work where he could, usually in mining towns, since this was what he knew best. Finally, in the United States, he settled in Pennsylvania where he worked in the coal mines. Soon, however, he began sailing again. He signed on merchant ships and sailed around the world, eventually returning to Pennsylvania. Later he sailed again and, when he was nineteen years old, he was on a British ship running the Union blockade of the southern ports. English ships took out bales of cotton and helped the Confederacy any way they could. With his ship unable to leave, Sam sat out the war between the states.[55]

Many people had mixed loyalties and, of course, some men did not want to fight in a war. A young lawyer bargained with Sam to take his place should he be called. He was to pay Sam $1,000. Fortunately, the lawyer was not called, so Sam was relieved of his surrogate duties. He nevertheless collected his fee in gold and returned to Pennsylvania when the war ceased. The

Civil War left many people destitute so they looked to the West as a possible salvation. Sam signed on a ship bound around Cape Horn for California where he planned to make his fortune in the gold fields. Archer and a man by the name of Frazer became partners and purchased four acres in San Francisco near the bay. They prospected in various parts of northern California for gold and coal. They spent one season working in the Mount Diablo coal mines. Back in San Francisco, Archer noted ships unloading coal from Coos Bay, Oregon. In 1869, Archer sold his San Francisco property and headed for Oregon.[56]

Sam Archer was a small man, only five feet seven inches in height. Hard labor in the mines and on sailing ships, however, developed broad shoulders and small hips. His small stature bothered him, and he was especially self-conscious of his small feet. Instead of wearing the size six shoes that fit his feet, he wore size eight. The long shoes soon turned up on the end and added to his pixy personality. Archer more than lived up to his appearance because he was always playing tricks and practical jokes, but he made friends easily.[57]

Arriving in Empire, Oregon in 1869, Archer moved into the local hotel. He was twenty-two. He easily hired on with Patrick Flanagan in the Newport Mine at Libby. Archer worked in a number of the local mines. He also prospected for gold on the beaches.[58]

Coos Bay was somewhat like his homeland of Scotland so Sam began to settle down. In 1872 he and Hiram T. Hall each built a cabin on land deeded under the Homestead Act. The Bay Area Hospital now stands at this location. He bought and sold other properties, primarily in the Libby area near the Newport Mine. During this time Sam met and fell in love with Margaret Monroe, who had come west with her father, Neil Monroe, in 1872. Sam Archer married Margaret Monroe in 1874. He was thirty and Margaret was fifteen. Margaret's children remember her as being very pretty but rather shy. She had long hair and wore it in braids hanging down her back except when attending social gatherings. Her abundance of hair gave her headaches. Sam was, of course, always the life of the party and a jovial story teller with many seafaring stories of adventure. Margaret knew

Samuel Archer with granddaughter, Mary Noone, in Goldfield, Nevada.

how to handle Sam and his cantankerous spirit. In 1875 Archer acquired 179 acres in the Bay City Area on the eastern side of Isthmus Slough. This became Sam and Margaret's home and the home that their children and future generations would know as The Red Barn.[59]

In the 1880's Bay City was a thriving community. Activity increased when Charles Merchant, manager of E. B. Dean and Company, moved the sawmill from downtown Marshfield to Bay City and combined his machinery with the Lobree mill. Merchant's crew moved the mill on barges to the new location. In those days the only way to get to Bay City was by boat. Most of the mill workers lived there in a bunkhouse and ate in the company cookhouse. Ellie Monroe was in charge of the cookhouse for a time.[60]

In the 1880s Eastside was not as large as Bay City. The ferry transported passengers from Marshfield, on Park Street, to

Timmerman's Point in Eastside. Many ranches and farms stood along the Coos River and Catching Inlet. The wagon road from Roseburg came in on the south of the Archer's property, and people built homes along this route. The "Bay City Society" brought many of them together for social occasions. *The Coast Mail* reported many meetings, pie socials, and dances in Bay City. Family gatherings were frequent. Often friends and relatives stayed several days at the Red Barn Ranch. Cousins played with cousins and listened to their parents' stories of the old times in Scotland and Kentucky.[61]

Margaret died when she was only thirty-two, so Neil and Elizabeth offered their support for the care of their grandchildren. The grandchildren were William, Neil, Martha, Grover, and Ellen.[62] Neil Monroe died at the age of 83 on February 2, 1902, and Elizabeth died two years later in 1904.[63]

In 1867 in Lanark, Scotland, William Archer's daughter, Rachael, married Forbes Thompson, and William thought then that he would like to go to America. He did not know how he could afford it, but through a mutual friend he got word to Sam of his needs. Because of the treatment from his father when he was very young, Sam was reluctant to help. He relented, however, and sent his father money to come to America. William Archer brought his anvil that he felt he could use in America. He went first to Canada where he visited his brothers in Collingwood, Ontario. He traveled to Illinois before heading west. He came to Oregon by way of California and soon acquired 160 acres adjoining Sam's property at Bay City. William set up a blacksmith shop across from the county road at Millington and spent much time there with his cronies, working and visiting.[64]

William Archer, also known as "Three-fingered Jack," suffered from a Carpal Tunnel Syndrome causing paralysis and severe pain in two of his fingers. One evening after Archer and one his cronies had a considerable number of drinks he laid his fingers on the anvil and had his friend chop them off to get rid of the problem.[65] William lived near his son, Sam, for 10 years before his death in 1880.[66]

Samuel Archer became restless after Margaret's death. He left his children with the Monroes and began traveling. He prospected for gold in Nevada and Alaska and worked in many places in the West before his death. In Alaska, Sam found it difficult to get work because his hair and beard had turned snow white. Potential employers thought he was too old to work, so Sam went to a barbershop and had his hair dyed red.[67]

Martha Archer married Wilkes Cavenaugh. They had three children: Ethel, Herman and Archer "Buck" Cavenaugh. Grover died at age five and Neil never married. Samuel Archer's many descendants remember him as "Grandpa Sam." These descendants are several families known in the area. They include the names of Archer, Cavenaugh, Wall, Menzenberg, Peterson, Earl, Mauer, Moody, Gress, Baird, Heaney, Aurdahl, Allen, Richardson, Fox, McCullough, Van Burger, Harrison, and Dyer.[68]

In 1910 Sam Archer went to Scotland to visit his sister, Rachael Thompson. Rachael's widowed daughter, Jean, had married a man named Forbes Brown. She had three children, two sons, Forbes and Adam, and a daughter, Eleanor. Sam brought Jean and her children to Coos Bay. Jean later married Edward Shepard. They provided a home for Sam when he was in the Coos Bay area. When Eleanor grew up, she married Charles Forrest. After Forrest died, she married Harold Grappe. Jean had studied to be a nurse in Glasgow. She rowed across the bay many a dark night in her duties. Later she had a motor boat that she used to go to homes on special cases. She spoke with a quiet Scottish brogue and when she agreed with a statement by anyone her answer was always a soft "aye" from deep in her throat. Jean (Thompson) Shepherd lived for 101 years.[69]

Sam continued his adventures and traveling until his death, January 3, 1935, in Goldfield, Nevada. He had been staying at the home of his daughter, Ellen. Ellen's husband, Jack Noone, was an undertaker. The Noones had a daughter named Mary.[70]

The Brynns. Another relative of the Archer family was Philip Brynn. In 1880 Brynn emigrated from Wales and settled in Illinois where he became a mining engineer. Within a few years he was able to send for his family in Wales. When they arrived they found Brynn was in jail on a "trumped up" charge. The mine owner thought that Brynn was going to quit when his family arrived. Apparently the mine owner had concocted a scheme of getting Brynn in debt to him so that he could keep him around. The wily Welshman was too much for the mine owner. Brynn, small and slight in stature, had his wife bring him in some grease. That night he took off all his clothes and greased himself thoroughly and forced himself through the bars and thus escaped.[71]

Mrs. Brynn was waiting outside with their belongings in a covered wagon. Brynn seated himself in the wagon and they headed for Arizona. The bouncing of the wagon caused the pain in his chest to give him considerable discomfort for the first couple of weeks of this journey. Philip had probably cracked a rib or two in his desperate escape between the bars. After weeks of traveling they finally reached Arizona where Philip found work in a mine. They became apprehensive about being found and taken back to Illinois; so they went to the mines in British Columbia. Philip found work on Vancouver Island, probably for the Vancouver Coal Mining and Land Company or perhaps for the Canadian coal baron, Robert Dunsmuir, in his mines at Cumberland or Ladysmith.[72]

A few months later the Brynns moved again. This time they found the mines at Coos Bay where Philip went to work in the Newport Mine. A few years later one of Brynn's daughters, Sarah, married William Archer, son of Sam and Margaret Archer. William Archer and Sarah Ann (Brynn) Archer had seven children, a daughter, Genevera (Archer) Richardson, and sons, William, Eldridge, Garnet, Harvey, Bruce, and Samuel.[73]

The connections of the Archer family to mining in the British Isles, their migration, and that of other relatives, to the United States, and their settlement on Coos Bay were representative of hundreds who came to the coal fields of southwestern Oregon. They and other found hard work but a good life in a new land.

ENDNOTES

1. John Eliot Allen and Ewart M. Baldwin, *Geology and Coal Resources of the Coos Bay Quadrangle, Oregon, Bulletin No. 27*, pp. 79-80.
2. Dr. Robert Macy, Interview with Dow Beckham, August 26, 1992.
3. John Eliot Allen and Ewart M. Baldwin, op. cit., p. 72.
4. Thurston Jacobson, Interview with Dow Beckham, June 25, 1992.
5. Allen and Baldwin, op. cit., p. 72.
6. Ibid., p. 57.
7. Ibid., p. 72.
8. Ibid., p. 78.
9. Ibid.
10. Ibid,
11. Ibid., p. 80.
12. Ibid., p. 67.
13. Ibid.
14. Ibid., p. 117.
15. George Stovall, Interview with Dow Beckham, July l4, 1992.
16. Emil R. Peterson and Alfred Powers, *A Century of Coos And Curry*, p. 392.
17. Allen and Baldwin, op. cit., p. 110.
18. Patterson Chaney of Santa Barbara, California, formerly from Coquille, Oregon, furnished the stock certificate of the Hardy Mine.
19. Victor West, The Hardy Mine, MS, December 23, 1992.
20. Allen and Baldwin, op. cit., p.110.
21. Dan Lynch, Interview with Dow Beckham, October l, 1992.
22. Allen and Baldwin, op. cit., pp. 110-111.
23. Ibid.
24. Ibid., p. 121.
25. Ibid., pp. 124-125.
26. Ibid., p. 128.
27. *Coos Bay Times*, October 11, 1909.
28. Allen and Baldwin, op. cit., p. 129.
29. Ibid., p. 136.
30. Andrew J. Davis, "Accounts and Correspondence, Coos Bay, Oregon," MS.
31. Orvil Dodge, *Pioneer History of Coos and Curry Counties, Oregon*, pp. 163-165.
32. Bunker Hill is a residential area south of Coos Bay between the outlets of Coal Bank Slough and Isthmus Slough.
33. Ibid.
34. Carol (Archer) Fox, Biography of Monroes, Archers and Brynns, MS, 1992.
35. Ibid., p.1.
36. Ibid.

37. Ibid.
38. Ibid.
39. Ibid., p. 2.
40. Ibid.
41. Ibid.
42. Ibid.
43. Emil R. Peterson and Alfred Powers, op. cit., p. 452.
44. Carol (Archer) Fox, op. cit., p. 2.
45. Ibid., p. 3.
46. Ibid.
47. Ibid.
48. Ibid.
49. Ibid., p. 4.
50. Ibid.
51. Ibid.
52. Ibid.
53. Ibid., p. 5.
54. Ibid.
55. Ibid.
56. Ibid., p. 6.
57. Ibid.
58. Ibid., p. 7
59. Ibid.
60. Ibid.
61. Ibid.
62. Ibid.
63. Ibid., p. 8.
64. Ibid.
65. *The Oregonian*, Section C 6, July 21, 1992.
66. Carol Fox, op. cit., p. 8.
67. Ibid.
68. Ibid.
69. Dr. Everett Curry and Anna Beckham, *First Baptist Church of Coos Bay, 1879-1979*, p. 38.
70. Carol Fox, op. cit., p. 8.
71. Ibid., 9.
72. Ibid.
73. Carol (Archer) Fox, 1992, Interview with Dow Beckham and permission to use family biography, March 10, 1992.

FIVE

ISTHMUS SLOUGH MINES

Southport Mine. In 1875 B. B. Jones opened a mine in T26S, R13W, for the Black Diamond Coal Company of San Francisco. The mine was about five miles south of Coos Bay on the west side of Isthmus Slough near the present day small animal facility. The company owned 2,700 acres that geologist J. S. Diller believed was part of the Beaver Hill coal bed. The Black Diamond Coal Company was controlled by Darius O. Mills and A. Hayward, both multimillionaires of San Francisco. By March of 1877 the miners had extended the tunnel 2,000 feet through the hill. Ships moored at the nearby dock and loaded out coal. They ran the tunnel in a northerly direction, bearing slightly to the west. A small village sprang up in connection with the mine. The owners mined about seven years before George Holt, who had bought the Newport Mine, paid them a $1,000 per month to discontinue. These payments lasted only a short time, as Holt became bankrupt.[1]

Thirty-Six years later, James Flanagan, son of Patrick Flanagan, leased 600 acres in this area from the owners, the Southport Land and Commercial Company. Flanagan mined coal there for several years quite successfully. In 1943 Oregon state geologists estimated that the operators took approximately 250,000 tons of coal from the mine. Since the miners had kept inadequate records, the geologists applied the 60% recovery figure for their estimate. The miners removed coal from about 55 acres. In 1937 James Flanagan told Engineer F. W. Libbey of the

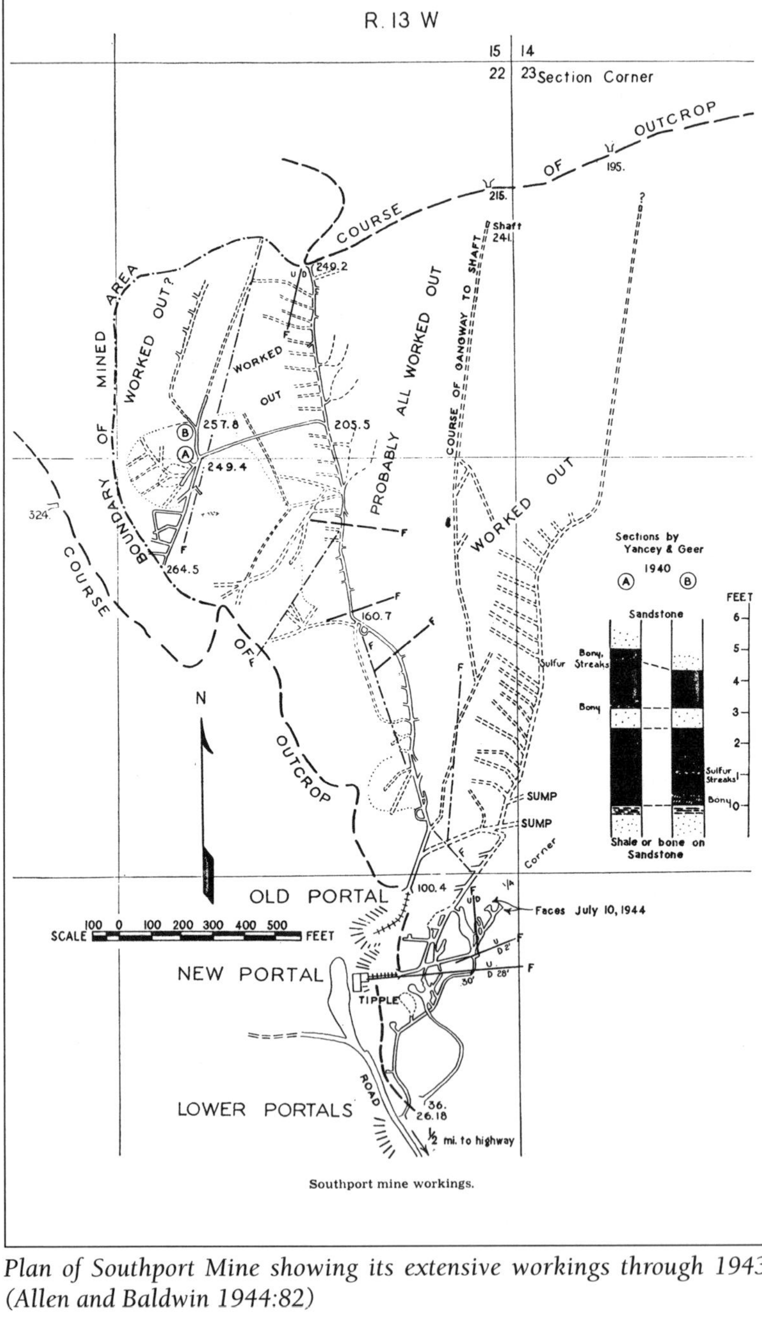

Southport mine workings.

Plan of Southport Mine showing its extensive workings through 1943. (Allen and Baldwin 1944:82)

Oregon State Department of Geology and Mineral Industries that he could mine over 200,000 tons of coal from above the entry level, in addition to the reserve available in the area that he was mining.[2] Flanagan held the lease until 1943 when he sold to Paul Murphy of Portland, Oregon. Murphy, developer of Westmoreland, Eastmoreland, Garthwick, and Lake Oswego, mounted no new work at Southport. During those years Flanagan shipped considerable coal to Portland over Southern Pacific's new railroad into the area. Smith-Power Logging Company also purchased coal from the Southport.[3]

The Southport bed was relatively level with just enough slope that the miners did not have to pump water. The down hill slope made easy pulling of the loaded cars. The miners took coal from "rooms" about thirty-five feet wide and relatively the same length. They left a pillar of coal in the center of each area for roof support. They excavated several gangways in the process of mining the coal. The gangways usually went at right angles to the tunnel.[4]

Miners in the Northwest at Coos Bay and in British Columbia found the mule an excellent animal for the laborious task of hauling coal. Every mine that used mules for haulage spawned a host of stories about the "doggone mules." More than mere beasts of burden, the mules assumed personalities and drivers respected them for their intelligence. In an area too low for the mule to stand upright, the animal dropped down on its knees to get through. Drivers taught some mules to push cars with their chests on very steep inclines. A mule pursued by a runaway car ducked into the first recess in the wall and let the runaway go clattering by. At quitting time the mule driver threw the harness off and turned the mules loose. Away they went at a fast trot hurrying to the stable and feed. Sometimes the young drivers grabbed hold of their tails and slid along the rail.[5]

Young boys learned all about the dangers and disagreeable aspects of mining. They also learned of the camaraderie among the miners and how much money they made. Most of the boys wanted to be diggers. A boy, tired of school, started making a little money by working at the mine. If he were willing to lie a little and add a couple of years to his age, he could start work at thirteen or fourteen. For the first year or two he worked at the pithead, picking tables for the diggers. If he were lucky, however, he would soon go underground. That was where the money was. He worked his way up and learned mining techniques. First, he ran a winch hoisting cars up steep inclines in some mines. Then he became a rope rider. The rope rider tended loaded cars on the way out of the mine.[6]

The mule and his driver took empty cars from the bottom of the last hoist and delivered them to the diggers. The diggers filled the cars and called for more. There was constant pressure on the driver to supply empty cars and take the filled ones out. A digger was only as good as the number of empty cars he could fill. The driver kept the cars coming, or the diggers would curse him. He in turn cursed fate or the "doggone mule" when the animal balked or when a heavy car derailed.[7]

Driving on haulage was tough and dangerous work. The driver had to be constantly alert, both to the dangers of the mine and the mule. He could be caught between moving cars or his hands could be squeezed between the top of the car and a low roof. Pushing the car the last few feet after the mule was unhitched was heavy work. The driver and the mule had to be a team. Despite the state of war existing between them, a truce was often necessary, especially when the driver's feeble light went out from a sudden jolt. Then both became dependent on the animal's remarkable sense of direction to lead them back to the stables in the blackness so intense a man could not see his own finger in front of his nose.[8]

Mules were smart and ornery. They responded to a certain man whose touch they trusted. Just as easily, they refused to work for him if it was past quitting time or the load was heavier than usual. When a mule took a stubborn streak or when he got mean, it could drive even the sanest and most compassionate man to tears of frustration. During these times, the driver administered a few well-placed blows across the mule's rear end. The mule, being the kind of creature he was, worked out revenge later.[9]

Ninety-four-year-old Homer Gilfillan, who lived most of his life in Coos County and worked in the coal mines when he was young had this to say:

I worked in the Southport for Jim Flanagan. I drove a mule for the mine on two separate occasions. Each time took from one to two years. After you had gone a short distance back into the tunnel you found a good slope developed. When the mules drew the loaded coal cars outside, with no fully developed brakes on the coal cars, the mule pulled the car hard with no interruptions until it would run of its own accord along the rail. There, you unhitched the mules and drove them in a space dug out of the tunnel wall, and the coal rolled on down the tracks alone. You braked the cars in several ways. The commonest was to push wooden sticks between the "A" rings of the coal car. This blocked the wheel

Homer Gilfillan, born in 1901, worked as a boy in the Southport and Delmar mines on Isthmus Slough prior to World War I. Gilfillan, until his death in 1994, was the oldest surviving coal miner in Coos County. (Beckham Collection)

but did let it slide down the rails. Then the mules were re-attached to the car and pulled on down to the bunker.

Each mine had its own peculiarities. In the Southport, the coal was dug and blasted out in blocks. Each miner loaded his own car, as nearly all miners had to do that, so they could finally determine how much coal they'd move out. On returning, the empties were brought to the mining area and the front wheels allowed to run off the tracks. Doing this blocked any possibility that the loaded car would escape and run down and out along the tracks. Oh, there's some mean mules, boy! They'd squirt you and some would bite. As you went by them they'd snap your arm if you weren't watchin'. And kick? They could kick I'm telling you. You'd harness them and they'd have a chain about six feet long with a hook on it to hook cars. They'd keep kickin' that chain until it hit you in the head. And they knew when it was quittin' time too. If you tried to drive them back for another load they'd balk and no way could you get them to go in again.[10]

George Welch gave this account of working the Southport Mine.

The coal vein was similar to the Newport mine though not so thick. It had one fire-clay mining seam and a good sandstone roof. The Southport had a number of owners. One was James H. Flanagan who leased the property about 1930 and worked the mine 'till 1944. I worked there in the winter of 1942-43. There were three miners, Tom Kelly and Andy Kelly, both old time miners. They were each over sixty years old and had both worked at Libby and Beaver Hill in the mines. Tom was the mule skinner. He had what we called a "spike team," two mules hitched one in front of the other. Tom and I would each load a car of coal, and then Tom would make up his "trip." He did not use lines but just drove by voice. He had a real muleskinner's vocabulary. He'd scream and curse and throw lumps of coal at the mules to get them going. Then they would go out of the mine on a dead run with the coal cars bouncing and rocking on the little rough narrow railroad track. Tom fastened an extra carbide miner's lamp on the head mule to light the way.[11]

Robert Pittam hauled coal for Fred McCullough from the Southport in the 1930s. Joe Beebe also ran the Southport during those years.[12]

Evor Rudberg worked at the Overland Mine between the head of Isthmus Slough and Beaver Slough. Mules were the miners' best friends and dragged tons of coal from the shafts. (Chard Collection)

A speculation and debacle, the ambitious investments at Henryville stood on the east side of Isthmus Slough about a half mile above the mouth of Davis Slough. The mine consumed $200,000 but produced no marketable coal. (Slattery Collection)

Henryville. The Henryville project became one of the most publicized mine failures in the Coos Bay coal field. California capitalists continued investing in Coos Bay coal fields before they finally decided scientific surveys were necessary. In 1873 Dr. C. A. Henry led a group of San Francisco capitalists and formed a company called the Northern Pacific Coal Company. They organized to develop 1,040 acres on Isthmus Slough. The company hoped to use its $2,000,000 capitalization for lumbering, transportation, and coal mining. In July, 1874, the corporation reorganized as the Coos Bay Oregon Coal Company to sell $5,000,000 in stock. That month a writer for the *Coos Bay News* paddled up the bay to visit the speculators' headquarters. He wrote:

> *It lies on Isthmus Slough, about half way between Coos City and the head of navigation. The whole ground is underlain with coal, as the croppings manifest themselves in every ravine. The frontage is bold and has sixteen feet depth of water at low tide. The distance is about twenty-five miles from the ocean and seventeen from this place [Empire]. It is just three weeks ago, since Dr. Henry and*

his 'fidus Achates,' S. D. Megeath, jumped ashore from the Messenger *and tramped it though the mud to their tent on the hillside. Since then a wonderful change has taken place.*[13]

Dr. Henry's transformation was to alter a recently logged hillside into an instantaneous frontier town. He hired Fox and Werden of Marshfield to place the piling for the new wharf. Almost immediately, he had carpenters plank it and build a two story warehouse and a store. He erected a boarding house with dining room on the main floor and bedrooms upstairs. He opened a blacksmith shop to prepare the coal cars. He built bunkers beside the slough and constructed a railroad track 1,375 feet from the mine portal to the dock. In the third week, Dr. Henry began construction of fourteen houses for his miners. In January of 1875, Henry established a post office with Steven D. Megeath as postmaster.[14] The geologist Goodyear, thoroughly acquainted with the Coos Bay coals, visited the Henryville Mine in March, 1877, and caustically wrote:

It is close to the waters edge and the situation would have been good enough if a mine had existed there. But there was never any reasonable evidence of any probability of the existence in this property of a pound of coal.

Dr. Henry founded his mine on a vein whose only richness was an abundance of bony and clayey partings. Also, the vein consisted of frequent dips and dives, bends and meanderings, and intrusive faulted materials. The costly steam engines for hoisting, the 900 feet of trestle, the bunkers of 1,000 ton capacity, the shaft and shoring, the coal cars, and the many buildings of Henryville had consumed $200,000. In May, 1876, the investors abandoned the site. Later, other speculators tried to develop the location. In 1877 miners dug 300 feet below sea level. Again, in 1885, William Phillips opened it. Phillips pumped it out and hired crews. A gas explosion that November injured three men and the operators closed the mine.[15] Matt Bowron served as watchman at Henryville for many years while the abandoned property deteriorated.[16]

Maxwell Mine. Some residents of Coos County remember the old bunkers that stood on the east side of Isthmus Slough near the mouth of Davis Slough. A pile of scrap iron, yet present in 1995, marks the location. James Wall opened the mine before 1900 and began mining coal. In 1902 the Crescent Coal Company, with W. H. Maxwell as superintendent, took over and operated until 1906. They took out 1,000 tons of coal according to Charles Sneddon, who worked there. Evor Rudberg also worked there and was familiar with the operation. These miners suggested that the market conditions and mining costs brought about failure of the project.[17] The 1944 survey indicated the six foot bed included a foot of partings and boney material. The miners drove the tunnel 500 feet into the mountain.[18]

Thomas Mine. In 1914 the Southern Pacific Company operated a mine across from the portal of the Southport Mine in the SE1/4, section 22, T26S, R13W. W. J. Thomas purchased and operated the mine on a small scale near the surface during 1944. The coal sloped 17 degrees south and 55 degrees east. Miners dug the tunnel 700 feet from the adit. The eleven foot bed contained only five feet of good coal. The miners who worked there in 1914 said the coal faulted quite drastically. Some people referred to the Thomas Mine as the New Southport.[19] Orville and George Chard, Evor and Ottel Rudberg, Jimmy Yates, and John Anderson and some of the Elliotts worked at the Southport and Thomas Mines.[20]

Utter City and Carbondale. Directly across the slough from Henryville on the west bank of Isthmus Slough, Coos County produced another ghost town, Utter City. In 1995 a few old piling showed the location about a half-mile south of the Highway 42 and 101 interchange.[21] Utter City had a hotel, stores, houses and a post office. At first, William Utter called the place Isthmus; Gilbert Hall served as the postmaster. February 11, 1875, the investors changed the name to Utter City. Zachery T. Siglin succeeded Hall as post-master, June 26, 1876.[22] William Utter and A. V. Ojeda, capitalists behind this venture, built a railroad

Above. *The* Marie, *owned and operated by George R. Chard, carried freight and passengers to the mines and farms along Isthmus Slough early in the twentieth century. (Chard Collection)*

Right. *Henry Sengstacken came to Utter City in the 1870s. Sengstacken eventually became one of the largest land owners in Coos County and served as mayor of Marshfield. (Beckham Collection)*

five miles from this deep water port to the south. They went over the divide to their mine called Carbondale. They built the small hamlets of Utter City and Carbondale entirely on the premise that coal could be brought profitably from the Carbondale Mine. Carbondale also became a ghost town when the venture failed.[23]

Utter City experienced some optimism when Henry Sengstacken came from San Francisco and established his business. He named his enterprise the Utter City Commercial Company. Sengstacken emigrated from Germany to San Francisco in 1865. Sengstacken attended and graduated from Heald's Business College. He worked a few months for his uncle, Claus Spreckels, the future sugar king and investor. Later, Spreckels invested in the Beaver Hill Mine soon after it opened.[24]

The Isthmus Transit Railroad (ITRR) was an important part of Utter's plans. He sent J. F. Dunham to San Francisco to obtain a locomotive. Dunham furnished the specifications for the engine and later named it the *Isabella*. Dunham shipped it to Coos Bay on the three-masted schooner *Emma Utter*. Dunham assembled the locomotive and became its first engineer. Travelers going to the Coquille Valley went on the Isthmus Transit Railroad from the head of Isthmus Slough across the divide to Coaledo. There, they took a rowboat down Beaver Slough to Freedom. The Freedom Hotel, on the Coquille River at the mouth of Beaver Slough, catered to travelers to and from the valley.

In 1874-75 Utter and Ojeda built Carbondale, a lively hamlet situated in a densely timbered canyon not far from Beaver Slough, a tributary of the Coquille River. Carbondale boasted two blacksmith shops, a store, and two hotels, one of the latter immodestly named the Bonanza. There were several comfortable residences. Utter and Ojeda opened the Carbondale Mine about three-fourth of a mile southeast of Coaledo near the corner of sections 14,15, 22, and 23, T27S, R13W.[25] Coaledo was on the north side of Beaver Creek where today's Highway 42 bridge crosses.

In the Carbondale Mine the lower tunnel ran south 450 feet. It cut a vein of coal six feet thick from gangways that ran east and west. The east gangway ran 1,100 feet on the six-foot vein, then diverged south 80 feet. There it cut a three and a half foot vein and extended another 200 feet. The west gangway extended through an unbroken six-foot vein 1,200 feet to the face. The operators considered the quality of the coal as excellent. The other tunnel was about 100 feet above the first one and ran south 200 feet. There it struck a three foot eight inch vein they named the Pelican. In 1875 Utter and Ojeda built bunkers at these mines with 500 ton capacity.[26] Although the company shipped 10,000 tons of coal in 1876, prospects for success were grim. The following year, in 1877, the Utter City-Carbondale enterprise failed. Utter tried to recoup some of the railroad costs by hauling freight and passengers. Sengstacken moved his business to Marshfield and became an astute businessman.[27] New owners moved the hotel form Utter City to Marshfield on a barge. Colonel J. D. Fry of San Francisco purchased the Carbondale Mine at sheriff's auction for $40,000. A year later William McCrindle bought the railroad.[28]

Delmar Mines. In June of 1907 C. A. Smith built the largest electrical sawmill in the world on Coos Bay. It began sawing lumber in January of 1908. The mill ran for over seventy-five years, through successive companies, cutting an average of 150 million board feet of lumber per year. The company established seven logging camps around the bay at first, but soon they found it necessary to reach back into the mountains for timber. C. A. Smith owned hundreds of acres of timber along the shores of Isthmus Slough. In 1905 the Southern Pacific Company bought the railroad built to Myrtle Point by Coos Bay, Roseburg, and Eastern Railroad Company. When Albert Powers started logging in the Isthmus Slough timber, he began to haul logs on the rail-

The sawmill of C.A. Smith Lumber Company stood at the mouth of Isthmus Slough and operated for over 70 years.

road. Powers used shortline roads as spurs to the mainline. The logging company burned coal in locomotives hauling logs to the large mill on Coos Bay. Smith-Powers Logging Company miners opened three tunnels at the head of Manning Gulch which branched off to the southwest near the head of Isthmus Slough.[29] The Smith-Powers Logging Company named the Manning Gulch (Delmar) area New Henryville. They used the mines from 1914 to 1916. The company purchased coal from James Flanagan's Southport Mine and also from the Noah and Delmar mines in Manning Gulch.[30]

Homer Gilfillan's father came to the area to work in the mines and soon sent for his family. Homer gave the following account:

> *Our family came into Coos Bay from Portland on the* S. S. Breakwater *in June of 1913. From Coos Bay we took a small gasoline launch, called the* Banshee *owned and operated by Captain Ross, to the headwaters of Isthmus Slough. There we went ashore at King's Landing. Father met us with a rowboat and rowed us across the slough to our new home. It was a small, but comfortable, three-roomed house and sat close by an old brick yard.*

Workmen installed the latest paper-making machinery in the C. A. Smith pulp mill on Isthmus Slough. The salt water process failed and the plant investment moved Smith closer to bankruptcy. (Stadden Collection)

My father worked in the nearby Delmar Mine. School wasn't due to begin until September. As children, we had few chores. Our family did not operate a farm as many in that day did. However, the lack of chores for me to do, rarely led me into problems. Instead of getting into mischief, I'd spend happy hours watching the operation of the coal mine. I'd snoop about in places where now-a-days, children would be run off. We'd be told, 'Our insurance doesn't cover accidents here. You kids get on out of here and keep out. We mean it.' But we failed to heed.

Despite their warnings we examined the tracks that went up to the mine; the coal bunkers into which the coal was dumped; on up to the mouth of the mine, and then back down the tracks again. We also went on about another half mile where a coal mine had been started earlier. It was known as the 'Ascot' Mine. In spite of family opposition, we did go into the abandoned 'Ascot' and looked around. We had no customary head lamps, so we couldn't go far.

The coal bunker for the Delmar Mines was almost at the end of the tracks from the mine, and was close to the

C.A. Smith erected a large, concrete pulp mill on Isthmus Slough to use wood products. The technology—founded on use of salt water—failed. The plant stood idle for decades. (Beckham Collection)

abandoned logging camp, opposite the cookhouse which had been used by Smith-Powers. Several of their buildings still stood nearby and were used by the miners. There were about 25 or 30 miners employed at going wages of $2.75 for a 10 hour shift for common labor. The miners themselves, however, were paid by the coal tonnage they produced. A good miner could mine two or three tons a day.

The coal was broken and freed at first by black powder. Dynamite came later. The coarse powder was rated by "F's". It could be double-F, triple-F, or four-F. The powder grains would vary in size from a half-grain of wheat to the size of cracked corn or small peas. They shaped the powder in small pellets. The black powder was shipped to the mine in tin cans with five pounds to each sealed container. Black powder was quite different from gunpowder. Not only was it more coarse, but it was also slow, very slow. This was preferred because the blast didn't crush the coal but rather broke it up very easily. The miners wanted to figuratively "push the coal out gently," as the miners would say. There would be a "puff" and the coal would be shoved out and away from the solid coal behind. The freed coal was then shoveled into the chute.

In using blasting powder, the miner first drilled a hole into the coal. A specially developed device was used to hold the drill bits. The drill was held firmly in a frame and by use of a crank the drill was slowly drilled into the face of the coal to the desired depth. Commonly, the depth drilled, would be five or six feet. The drilling device would be removed, the hole was packed with black powder, with a fuse attached. The manner in which the powder was used was unusual in itself. The miners took newspapers and rolled them about a shovel handle to make a tube about 18 inches long. Several thicknesses of paper were used, after which one rolled end would be carefully crimped and closed off. The newspaper tube would be slipped from the shovel handle, and loaded full with powder; a black fuse was pushed into the powder in the open end, and was then tightly tied off. This powder tube was pushed into the newly drilled hole, with the powder fuse hanging out the hole. The balance of the hole was not filled by powder but was packed tightly with 'dirt,' but never coal. The miners called the layers of clay that they found in the band of coal 'dirt'.

The black powder blast was set off twice each day, usually. The frequency of blasting was based on the presence or absence of coal to be shoveled into the chute. Black powder was used between 1913 and 1915. Later power drills or mining machines were used. When they blasted, all the miners throughout the mine came together in the clear, in the area of an air shaft, well removed from the blasting. They tried to schedule the blast just at lunch time so that while they were eating, the slopes and the gangways would have time to clear of the dust and smoke.

Good air is very essential to a mine. The Delmar Mine I knew and worked in had a very large air shaft back up on the hill, possibly three hundred feet from the mine's entrance. The Delmar Mine had but one air shaft. However, air shafts are absolutely essential to a mine. Depending on the mine and its nature, more than one shaft might be dug. Mines also had cross air shafts, dug from one shaft to another directly through the coal itself. All air shafts had an average diameter of three feet. Large, power driven fans were placed to force adequate fresh air in and bad air out of the mine and reduce to a minimum the coal dust breathed by the miners. In 1914, the mine was equipped with electricity and a very large fan was placed at the top of the outside air shaft. This sucked out fumes and coal dust and forced fresh air to be drawn in through the mine entrance.

The mine surveyor was a very important person. His planning developed the best course to follow in removing the coal, as well as assuring the safety of the miners. In a small mine such as ours the mine surveyor would come on request. He'd then lay out a program that could take a few days or weeks to complete.

I remember three explosions at the Delmar Mine. There was one man killed and perhaps three or four injured rather seriously by flames from the gas explosion. Being a small

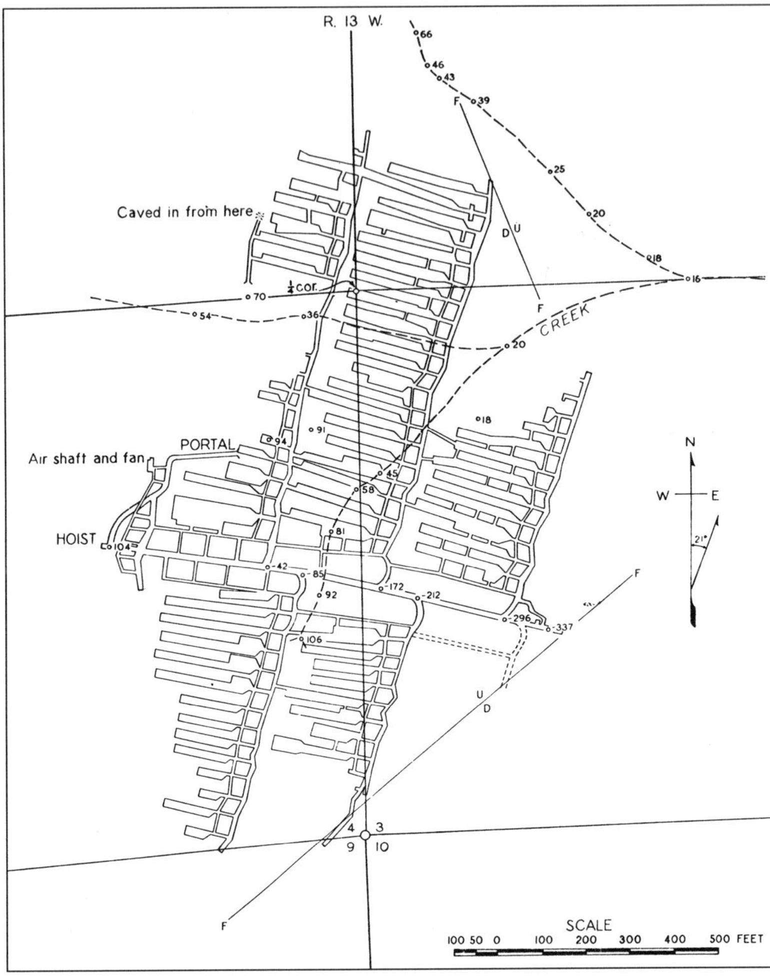

Plan of 1915-16 of the Delmar Mines as prepared by H. G. Butler and Lancelot A. Whereat. (Allen and Baldwin 1944:37)

mine the gas rarely went throughout the mine, but collected in a pocket while coal was being removed. The coal gas lies naturally in and among the veins of coal itself. As the coal is mined, the gas would escape into the tunnel in small quantities. However, should a large volume of gas collect over a weekend when the ventilating fans were not running, there could be a very disastrous explosion after the mine was opened and in use.

If miners wore a Davis safety lamp, the protected flame in the lamp would not cause an explosion. This safety lamp flame would change its color if gas were present. If so, the miner made haste to get out. At times the Davis lamp wick flame would also flicker. A small, usually harmless explosion within the heavy glass cylinder holding the flaming wick in the Davis lamp was possible, but it would not cause a general explosion.

As empty areas developed throughout the mine, when the coal and dirt were disposed of, the miners set timbers to support the mine roof. The timbers were commonly made from logs felled on the place. The logs would be split into poles about six or eight inches in cross sections. The miner had available to him a handsaw, wedge and lumber to make wedges and caps for the top of the timber. He'd strike the roof with his pick. If it had a sharp ring, it was secure. If the sound was hollow, he'd immediately set a pole tightly wedged into the roof above.

Mines in those days had their safety rules. However, in small mines like we had in Coos County, the rules were not always strictly enforced. Stretchers were available to carry anyone injured out of the mine. If there was an accident in the mine, and it was in the under-room or the gangway where the men were working, a hoist would let the empty coal cars down into the mine. The injured miner would be loaded into a car and the hoist would pull the car and patient to the surface.

I recall, vividly, the filling of the cars by the miners down in the mine. Each miner would be allotted two or three cars. Each car would hold about one and a half tons of coal. The men were paid a dollar a ton. A miner who got out five tons would thus get five dollars. Laboring men, other than miners, in the Coos Bay area got $2.75 a day in those days. The daily production of the Delmar Mine ran from thirty to forty tons. The mine could produce fifty tons if the miners were employed in sufficient numbers. In addition to the miners, the Delmar Mine also had a mule driver for the coal cars, an engineer who operated the double drum *small steam donkey engine; a blacksmith, a carpenter, the weigh man who weighed the coal, as well as, a full railroad crew. Not all the employees were engaged in digging coal.*

Our coal bunker at the Delmar Mine was about 50 feet by 50 feet and 60 to 75 feet high. It was built on a slope. The loaded car was pulled up onto the top deck of the bunker. A man using a large, heavy hammer knocked the side of the coal car loose at the bottom. In so doing all the coal in the car poured out into the tipple as the car was upset to about a 45 degree angle. The car was held on the tracks when emptied by a clever device which did not allow the car to go clear over on its side when the dump man pulled the lever. The cars were lined inside with heavy sheet metal which let the coal fall freely from the car. Once it had been emptied, the lever was reversed and the car again stood properly on the tracks. The coal cars were built and kept in repair by a blacksmith and carpenter. The cars were built for heavy duty. They were constructed of four inch lumber on the bottom and two inch lumber on the sides, all reinforced by angle iron and covered on the inside by sheet metal.

A small locomotive engine would pull or push empty coal gondolas alongside the coal bunker where they could be loaded by gravity from the bunker chutes above. After being loaded, the loaded cars were placed on a side spur which had been for the old log dump. From there the coal was taken to Coos Bay for shipment or local use.[31]

Overland Mine. George Hiram Chard, a developer of the Overland Mine, was the son of an Englishman, George Chard, who came to Coos Bay where he worked as a logger and salesman of merchandise. George H. Chard, born in 1883, married Clara Catherine Bosco. The Chards lived in the Delmar area and raised six children: Wanda, Leola, Lucille, George Raymond, Clayton, and Dorothy (Chard) Barton. When he was young, George H. Chard worked in the Southport and Thomas mines. He also operated a small boat, the *Marie*, on which he carried freight and passengers. The boat served the communities along Isth-

The Overland Mine operated from June, 1932, until 1944 near Coaledo in the Beaver Slough watershed. Miners included from the left Evor Rudberg, George Chard, John Anderson, and Ottel Rudberg. (Chard Collection)

mus Slough. A brother, James O. "Ole" Chard, also worked in the Southport and Thomas mines.[32]

During the Great Depression of the 1930s, George R. Chard, John Anderson, and Evor and Ottel Rudberg were like many others needing income. Times were difficult. They decided to start selling coal. They had access to some property one and one-eighth mile from the highway at the railroad overpass near Coaledo. On this property they knew of an outcropping of coal as a possible prospect. They moved in what little equipment they could get together and went to work. They opened the mine in the NE1/4, section 9, T27S, R13W, in Dunham Gulch, about 10 miles south of Coos Bay.[33]

The partners started in June of 1932. The Oregon Department of Geology and Mineral Industries estimated their total production during those years as 20,000 tons, some 2,300 tons a year. They actually had two mines, one on the north side of the gulch and one on the south side. They established a tunnel on the north side 400 feet long with rooms from 50 to 180 feet up the dip. On the south side, they dug a slope 700 feet long with two gangways at 400 and 700 feet. They made the lower gangway 2,000 feet long and about 140 feet below sea level. The coal bed averaged seven feet in thickness and had impurities that had to be separated in the mining process.

Eventually, six partners worked in the mine. They used a diesel-powered electric plant for lighting and pumping air and water. They hauled the coal out to the tipple with mules. They sold coal locally at $6.00 per ton for public buildings and a few households. The miners also sold to ships in the harbor at Coos Bay and to the U. S. Navy base in North Bend. Harold Anderson made up the blasting dummies and George Raymond Chard worked as a rope rider. Anderson wrapped the sheets of paper around a stick to form a tube that he filed with powder and inserted a fuse. The rope rider, of course, was the person who came out with the cars loaded with coal, to make sure that it arrived on the surface as intended. Another George R. Chard, a young boy, was the rope rider who tended the trips of coal out of the mine.[34] In 1944 the miners encountered serious faulting that made the labor too expensive. That and marketing problems caused them to stop working the mine. The Overland Mine was a part of the same bed of coal as the Beaver Hill Mines.[35]

Martin Mine. In 1937 Engineer F. W. Libbey called this mine the Thomas Mine. John Eliot Allen and Ewart M. Baldwin, in their 1943-44 survey, called it the Martin Mine. They referred to the mine across the gulch south of the Southport Mine as the Thomas Mine. Libbey gave the following report:[36]

> *Mr. Zeph Thomas has about 300 acres leased in section 9, T27 S, R13W, and is now mining in the SE1/4 of this section. The mine is reached by going south from Marshfield on Highway 101 for a distance of 10.5 miles to the north end of the town of Coaledo, and then turning west and northwest on a rough dirt road for a distance of one mile to the mine bunker.*
>
> *The coal is mined through an adit, driven in a northerly direction on the same Beaver Hill bed that is being mined at the Overland property. The Thomas mine was opened in 1936 and only a small tonnage has been removed. Present output is 5 or 6 tons a day. The room being worked*

at present is about 500 feet from the portal, and has been driven up the dip at an angle of about 35 degrees for about 30 feet. The coal is drawn through a chute into a one-ton car and hauled to the tipple at the bunker by one mule over a 36-inch gauge track laid with 30-pound rails.

At this working the bed has a section as follows: Sandstone footwall; Bone coal, 8-10 inches; Coal 20 inches; Sandy clay, 9-10 inches; Coal, 36 inches; Sandy clay, 8 inches; Coal, 12-16 inches; left for roof support, Sandy shale hanging wall.

Both benches are mined with a pick by undercutting and removing the clay seams. Stulls are used fairly close together in the room as the roof is not strong. The coal is washed over screens by a hose at the bunker. Three men are employed. Mr. Thomas delivers the coal by truck; chiefly at Marshfield. The property is one mile from the railroad at Coaledo. According to Diller's mapping, this whole quarter-section is underlain by coal, all above a depth of 2,000 feet.[37]

In 1940 H. F. Yancey and M. F. Geer, of the same department, gave a similar report but stated the miners had tunneled 600 feet into the seam. They also indicated that the mine produced an average of ten tons per day in 1938.[38] Allen and Baldwin reported that Zeph Thomas operated the mine from 1936 to 1941.[39]

Isthmus Slough mines provided employment and brought in capital to the Coos Bay area for over sixty years. Even the mine failures—beginning with the $75,000 investment of Andrew J. Davis and that of Dr. Henry with hundreds of thousands of dollars—helped the local people. The tremendous expenditure of Utter and Ojeda kept the local economy stirring for almost a decade. In 1890 when Graham began building the railroad through Coquille to Myrtle Point, the future of the area indeed seemed bright. Although the railroad was not finished to Roseburg, the Southern Pacific Company purchased the line and took over the Beaver Hill Mine for continued infusion of outside capital into the area.

ENDNOTES

1. Emil R. Peterson and Alfred Powers, *A Century of Coos and Curry*, p. 398. Peterson and Powers said that the mine opened in 1877. Allen and Baldwin indicated 1875 as the date.
2. John EliotAllen and Ewart M. Baldwin, *Geology and Coal Resources of the Coos Bay Quadrangle*, p. 80. Allen and Baldwin indicated that the Southport Land and Commerical Company owed the Southport Mine when James Flanagan ran it.
3. Emil R. Peterson and Alfred Powers, op. cit., p. 339.
4. John Eliot Allen and Ewart M. Baldwin, op. cit., p. 80.
5. Lynne Bowen, *Boss Whistle*, p.35
6. Ibid.
7. Ibid., p. 36.
8. Ibid.
9. Ibid.
10. Homer Gilfillan, Remarks to Coos County Historical Society, February 23, 1979, (typed by Dr. Raymond McKeown). Interview with Dow Beckham, March, 1992.
11. George Welch, Coos County Coal Mines, MS, p. 5.
12. Jack Beebe, Interview with Dow Beckham, September 14, 1992.
13. Emil R. Peterson and Alfred Powers, op. cit., p. 137.
14. Ibid.
15. Stephen Dow Beckham, Coos Bay *The Pioneer Period 1851-1890*, p. 25.
16. Emil R. Peterson and Allfred Powers, op. cit., p. 393.
17. John Eliot Allen and Ewart M. Baldwin, op. cit., p. 84.
18. Ibid., p. 84.
19. Ibid., p. 83.
20. George Chard, Interview with Dow Beckham, August 15, 1993.
21. Emil R. Peterson and Alfred Powers gave the location of Utter city as one mile south of Henryville. They also stated that it was five miles by the rail line to the Carbondale Mine. The Carbondale Mine was almost a mile southwest of Coaledo. Coaledo is just to the north of the bridge across Beaver Creek on Highway 42. Careful measurements show that the piling west of Henryville was the site of Utter City. The huge coal bunker in Maning Gulch was called New Henryville. This may have confused Peterson and Powers., op. cit., p. 393.
22. Emil R. Peterson and Alfred Powers, op. cit., p. 138.
23. Ibid., p. 392.
24. Ibid., p. 148.
25. Orvil Dodge, *Pioneer History of Coos and Curry Counties*, Oregon, p. 427.
26. Ibid., p. 24.

27. Emil R. Peterson and Albert Powers, op. cit., p. 392.
28. Stephen Dow Beckham, op. cit., p. 24.
29. Dow Beckham, *Swift Flows the River,* p. 71.
30. John Eliot Allen and Ewart M. Baldwin, op. cit., p. 86.
31. Homer Gilfillan gave permission to use his manuscript March, 1992.
32. George Chard, Interview with Dow Beckham, August 15, 1994.
33. Ibid.
34. Ibid.
35. John Eliot Allen and Ewart M. Baldwin, op. cit., p. 88.
36. The community of Coaledo is not usually referred to as a town.
37. F. W. Libbey, Oregon State Department of Geology and Mining Industries, Bulletin No. 2, *Progress Report on Coos Bay Coal Field,* p. 6.
38. John Eliot Allen and Ewart M. Baldwin, op. cit., p. 90.
39. Ibid., p. 90.

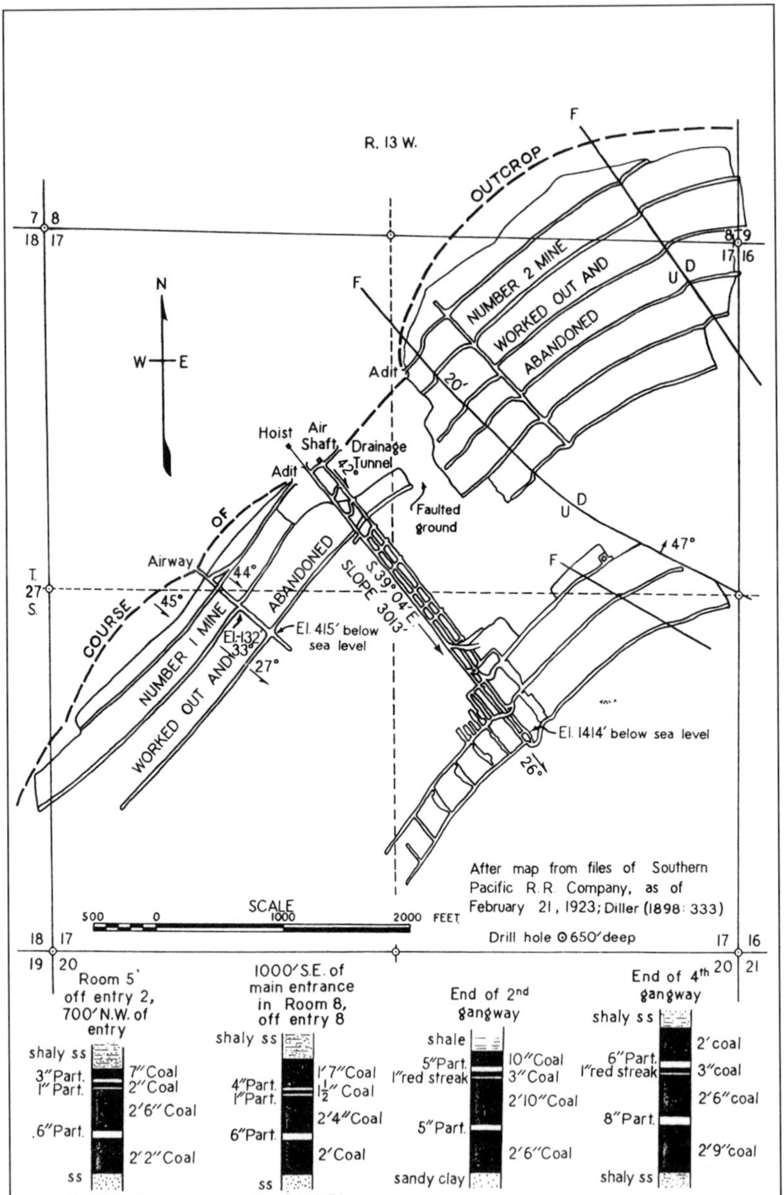

In 1923 officials of the Southern Pacific Railroad drew a detailed map of workings in the Beaver Hill Mine. By the 1920s, the shafts extended 1,414 feet below sea level and necessitated extensive pumping to remove water. (Allen and Baldwin 1944:93)

SIX

BEAVER HILL MINES

Each mine had its own peculiarities, from the small family production to the Beaver Hill Mine, the largest in the county. At the Beaver Hill Mine, when the men seated themselves in the cars for the trip to the depths, the foreman gave the signal and down they went. They dropped so fast that if a miner closed his eyes, he felt like he was floating. At the bottom, some 3,000 feet into the earth, the cars came to a shuddering halt. The shift bosses urged the men to hurry and not waste time. The shift being replaced climbed aboard the cars for the trip to sunlight. Others hurried to their work places so that they could get the coal moving again.[1]

At the bottom of the brightly lit shaft, 1,300 feet below sea level, the scene bustled with activity. Coal-laden cars clattered over the tracks, metal wheels on metal rails. Bosses shouted orders and marshaled their crews for action. They urged new employees to move into their places. The men disappeared into the labyrinth of the mine, some on empty cars and some on foot. The face of the coal echoed the noise of the machinery and occasionally shuddered from the deep "wump" of the explosions that pushed out the coal.

Just as the fisherman knows the sea and the logger knows the woods, coal miners knew the underground intimately. They knew the coal in all its variations, the rock supporting it, and the way the beds contained clay and useless material in the seams of coal. They knew a solid roof from a friable one and why both

Miners at Beaver Hill about 1900 included an African-American, second from left, and three boys in the center foreground. (Slattery Collection)

could be dangerous. Experienced miners acquired a sense of direction and distance in the maze of tunnels. The sounds the mine made were familiar. They could detect the subtle difference between what was merely the roof settling, or the timbers creaking, and unusual situations that warned of danger.

The miner knew the mine could kill him if he let down his guard. Like a fighter who respected his opponent, he kept a wary eye out for the lethal punch and, most of the time, was able to counter it. He knew that the mine could kill with the fall of a rock, or coal, or it could suddenly flood with water, or a runaway car could bash him the dark. He knew the deadly gas was a constant threat and that the slightest spark could cause an explosion. The ventilation system of the mine, designed to drive away the gases, worked as the miner's best defense. He felt secure when the fresh air sucked by the massive fan swept through the slopes and gangways.

At the working face, the small lamps on the miners' caps provided the only lights. These gave feeble light in the vast and deep darkness. Despite the sweep of the ventilation system, the air choked with fumes from the blasting made breathing difficult. Men coughed and cursed. They drilled and dug, and heaved heavy timbers. They worked on their feet, on their knees, and on their bellies—digger and back-hand, chunker and mucker, driver and rope rider. For each of the hundreds of men who worked in the Beaver Hill Mines, the experience was different. Some, in spite of the sweating and confinement, found the mine hospitable. Its darkness and even temperature reassured them. The damp and the danger, they saw as unpleasant but necessary in the pursuit of a living. Some who came, however, saw the realities of the mine and left, never to return.

For most men mining was a matter of adjustment. The young winch boy or rope rider overcame his fears and became a seasoned underground worker. The process of adjustment required an iron nerve, a determination never to give in to fear. The novice soon developed the ability to use the mine to his own advantage.

The lack of light accentuated the sounds, both the useful and the dangerous ones. Most miners never let the darkness bother them, even when their head lamps went out, as often happened. On those occasions, the miner usually called to someone nearby, or the man followed the car track with his foot until he came to other miners. Miners did not necessarily agree about the danger and drawbacks in the mine. Any group of miners would have a range of opinions. For example, some said the smells in a mine had an odor of gas mixed with stagnant water. Others laughed and said it was like strawberry jam and peanut butter sandwiches.[2]

As for the noise in the mine, some said it was too noisy to talk. Yet they agreed that they could hear the subtle creaks and rumbles that warned of danger. Some never thought that the mine was dangerous, only that people made it dangerous. The experienced miner rationalized that working in the mine was no different than going into a field and picking potatoes, or husking corn or pitching hay. Seasoned miners did not think of the mine as a dark and eerie place. They grew up with the atmosphere of the mine and thought accordingly. In those days, ranching, mining, logging or working in a lumber mill were the main occupations.[3] In southwestern Oregon much of this work was largely seasonal, until C. A. Smith and Al Powers came in 1907.

Things changed somewhat with their building of C. A. Smith's large sawmill on Coos Bay. The sawmills provided steady work, but not everyone could work in the mills.

Lumbering and mining attracted a number of racial groups to Coos County. Racial discrimination usually had economic roots. Established workers often looked questionably on the newest group, whether Italian, Japanese, Chinese, Finns, Swedes, Blacks, Mexicans, Irishmen or Scotsmen. They considered them a threat. They presented an unusual appearance, often because of color, dress and language. They worked for lower wages and sometimes as strike breakers. They looked different; they ate peculiar foods; and their talk sounded like gibberish.

For some racial groups, the hatred died hard. The American West discriminated against Orientals until long after they were no longer considered an economic threat.[4] Finns and other Scandinavians lived near each other and preferred the woods and mills to the mines. The Italians also suffered similar attitudes from the established workers. The 1900 census recorded places of birth for the diverse population of Beaver Hill: Oregon (32), Washington (5), Illinois (2), California (3), Italy, (2), Scotland (2) and England (2). A number of other states made up a total roll of 152. By 1910, the census showed 17 Japanese, 2 Chinese, 11 Italians, 11 Austrians and the balance from around the United States with many of them having foreign-born parents. The 1910 Beaver Hill census totaled 149. When C. A. Smith built his large mill on Coos Bay, many Scandinavians came from Minnesota to work for the lumber company. He also opened up job opportunities for those who lived in the area. Some miners moved to logging and lumbering. The Southern Pacific Company then brought in a contingent of Japanese to keep the mine going at Beaver Hill.[5]

Although the racial groups tended to live together and the community experienced some prejudice, down in the mine most racial distinctions became blurred. Some of the newcomers were single men who lived mostly in the hotel and ate in the huge cookhouse provided by the company.[6] Everyone worked hard and was entitled to respect regardless of his origins, though this not always accorded.

Like men everywhere who work under difficult conditions, a camaraderie developed which defied racial discrimination. Down in the mine they were all friends. Outside, maybe not, but down in the mine they treated each other as equals. It did not matter if they disliked or sensed racial differences on the outside. If someone asked for a hand down in the mine, he knew he would get help, even if they were not friendly on a personal basis. Miners kept a cooperative spirit of personal regard for other men and their safety. If a miner sensed something was wrong, or did not see or hear the other miners for a while, he stopped work to look. They were very helpful to one another. When doing heavy work, such as putting up timbers or stringers, they called on the other men to help. When a miner left his work, he left the basis of his earnings. Most of the time the miner was paid according to the amount of coal he dug.[7]

Illuminating the dark tunnels of a mine was only one reason for the presence of lamps. The miner used his lamp to test for gas. There were several types of gas in the mine that could explode if exposed to open flame or even a spark. They tested for gas, cleared it, and avoided an open flame or spark. For hundreds of years miners worked with the open flame lamps to light their work-place. Danger of an explosion was a constant threat. They used whale oil or fish oil, either of which had a terrible odor. Some of the little pitlamps had cotton wicks in the oil and some of them had a sort of a double spout. They wore them on their caps and quite often spilled the oil on themselves. Sometimes the small holes in the rib (the side of the tunnel) leaked gas into a pocket of water. For amusement, the miner held his lamp in the water and watched it crack and fizzle. He knew there was not enough gas to blow.

Overleaf. *Their faces begrimed with coal dust, miners from Beaver Hill emerged in 1916 from the shafts which extended 1,414 feet below sea level. The men held their tin lunchpails which provided a meal during their long shifts deep beneath Coos County. Lancelot A. Whereat (far right), a civil engineer born in England, worked for the Southern Pacific Railroad at Beaver Hill from the late 1890s until about 1916. Whereat ultimately served as mine manager and later labored in Washington, D.C., as a lobbyist for the SP. (Whereat Collection)*

Beaver Hill Coa

The gas must mix with a lot of air to explode. Using an open flame in a coal mine was dangerous.[8]

The carbide lamp was another light source used for a long time in the mines. It was much safer than the open flame lamp. The carbide lamp produced light when water dripped on carbide (a compound of carbon and calcium) making acetylene gas. The miner ignited the gas with a flint in an enclosed, glass-fronted chamber. He usually attached the carbide lamp to the soft peak of his cap.[9]

The safety lamp became one of the most useful tools of the miner. Its light helped, but detecting the presence of gas became very significant. No matter how explosive the gas was, black damp, carbon dioxide, or anything, the safety lamp would not ignite it. The flame was locked inside a chamber. Although mines installed electric lights in the second decade of the twentieth century, miners continued to use the safety lamp to test for gas. When the miner got to his work-place he turned his safety lamp flame down low and held it up near the ceiling. If fire-damp was present, a blue halo or "cap" appeared around the yellow flame. The first electric lamps worked on a wet-cell battery. If a miner were not standing erect, the battery acid leaked out and burned his back. These lamps were unsatisfactory.[10]

Fire-damp, as the miners called it, was a mixture of air and methane. The coal seams imprisoned methane within the coal, as well as above and below the coal seam. Usually the gas escaped slowly. Sometimes under pressure, it hissed like a jet.

Miners applied the word "damp" to any dangerous gas, and there were other types besides fire-damp. Black-damp, sometimes called choke-damp, was the most lethal. It killed by explosion or by asphyxiation. Black-damp was a mixture of carbon dioxide with either air or nitrogen. It leaked off the coal and blew up with a naked flame. Sometimes it hovered near the floor and caused a victim to become sleepy or die for lack of oxygen. In the aftermath of a mine fire or a fire-damp explosion, one of the dangers was the after-damp. In the first hours after an explosion, rescuers could not give assistance to their fellow workers until they cleared the mine of the deadly gas with its high concentration of carbon monoxide.[11]

The ventilation system of the mine was most important and had to be carefully designed and maintained. Miners carved tunnels and crosscuts to facilitate the flow of fresh air. A large fan pulled air into the mine and blew throughout the mine in one direction past every working face. The flow of air swept all types of "damps" out the return shaft. To guide the air into each working area and then back out, workers hung brattice boards. The brattice board or canvas curtains directed the flow of air. The "brattice" man was an important employee. He tended the brattice boards or curtains and constructed special ones when a pocket of gas needed attention.

"Stoppings" were another indispensable part of the ventilation system. "Heavy Stoppings" were constructed of two-inch, timber. Miners built permanent barriers with rocks or bricks to block off old workings where gas collected. They placed swinging doors for some stoppings in cross-cuts. The doors deflected the flow of air and kept it going in one direction. Trappers opened the stoppings briefly to allow the passage of a line of cars hauling coal or miners. Some mines employed young boys called "trappers" whose sole job was to sit in a small hole in the wall and swing the stopping back just long enough for the "trip" to pass through.[12]

These realities—depth, danger, darkness, and the prospect of death—were present in all Coos County mines, but especially in the vast and deep excavations at Beaver Hill.

In 1894 R. A. Graham opened the Beaver Hill Mine. Graham came to Coos County from California in 1891. He had grandiose plans to build a railroad from Coos Bay through Coquille and Myrtle Point to Roseburg. Graham called his enterprise the Coos Bay, Roseburg and Eastern Railroad. Graham solicited property owners who were to benefit from the railroad to pledge money for the endeavor. A number of people thought their property values would double. Graham laid only five miles of track the first year. Graham spent $100,000 of his own money and borrowed a half million dollars from J. A. Spreckels and Brothers of San Francisco. By September of 1893 the investors reached Myrtle Point with the railroad. The financing dwindled so they discontinued construction that year. Graham and his backers attempted to recoup some of their expenses by turning to coal mining at Beaver Hill. They built about a mile and a half of railroad in a spur across Beaver Creek and up Caulfield Slough to the mine opening.[13]

In January of 1895, Graham, desperate to find cheap labor, sent a representative to West Virginia to hire experienced miners. Eleven blacks (six of them with wives) and five or six whites responded on the promise they would receive $5 to $7-a-day wages, be furnished good living quarters, and have good working conditions. When they arrived at Roseburg, they were told they would have to walk the rest of the way. At the mine they found no housing. Wages were ninety cents per day and they had to wade through deep mud to get to their jobs. It was a cold and rainy winter and they had to live in box-cars until they could build decent houses. When they complained to Graham, he ordered them to pay their board and lodging and leave. The disgruntled and dismayed group walked ten miles to Marshfield. The miners at Libby took up

The depot and shops of the Coos Bay, Roseburg & Eastern Railroad, 1896, stood adjacent to the tracks leading to the Beaver Hill Mine bunker on the Marshfield waterfront. Central School behind the bunker overlooked the town. (Stadden Collection)

a collection and bought them passage on the next boat to San Francisco.[14]

The Beaver Hill Mine owners installed extensive surface improvements and built a large bunker on the Coos Bay for loading ships. Graham, however, was so deeply in debt to Spreckels that the absentee capitalists had no recourse but to force him out. Claus Spreckels came up from San Francisco and visited the mine. He placed William S. Chandler, an experienced miner, in charge. Chandler abandoned Graham's works and opened another slope, but used the same surface facilities.[15]

Other than a brief time about 1900, when the Southern Pacific Company bought out Spreckels, the mines at Beaver Hill operated until 1923. Production at times exceeded the company's ability to ship out the coal. A news item in the *Coos Bay News*, February 5, 1896, stated:

> *The Beaver Hill mine was shut down for a few days lately, the bunker being full. The* Blanchard *took coal, Friday, and the* Omer *relieved the bunkers to the extent of 600 tons,*

A steam schooner laden with coal from the Newport Mine bunker departed from the mouth of Coal Bank Slough. Upper Coos Bay prior to dredging and filling possessed a large tidal surface. (Stadden Collection)

The Coos Bay Coal and Fuel Company, owned and operated by Hugo Quist in the late 1910s and early 1920s, stored coal from the Delmar Mines in the New Henryville Bunker at Manning Gulch.

The Beaver Hill Mine tipple about 1921 overlooked the rails where crews loaded coal for shipment to Coos Bay.

Sunday, enabling the miners to resume work. Twelve more dump cars for Beaver Hill were forwarded from the car shops to the mine last week, and carpenters are now employed in building 25 ton cars, used in bringing coal from the mine to the depot bunker.

The Newport Mine bunker stood near the mouth of Coalbank Slough. The railroad crossed under the Newport Mine track to reach the Beaver Hill bunker that stood near the intersection of Elrod and Broadway in the City of Coos Bay. For a time, F. S. Samuels, secretary for the Spreckels Brothers, managed the Beaver Hill Mine. In April, 1896, the *Coos Bay News* reported: "More cottages are going up at Beaver Hill. A hotel will be built at that place in the near future." Social activities of the Beaver Hill community also gave an aura of sophistication as noted in the *Coos Bay News*, April 1, 1896:

First entertainment of group given on Monday evening in the enterprising young city. Piano forte solo, "The Sweet Long Ago," *by Miss Clara Davis, and F. A. Holdsworth, tenor, with* "At the Ferry" *Mr. J. K. L. Parker recited "Jack's Religion," guitar duet, etc. Very long program then a ball by the Foresters.*

These and other news item indicated that the Beaver Hill project was finally becoming a major activity in the county.[16]

Three distinct mines, known as the Beaver Hill Mine, over the years used the same surface facilities. The Klondike Mine nearby, also opened into the same bed of coal. The Beaver Hill Mine opened about a quarter of a mile north and west of the center of section 17, T27S, R13W, near the upper end of the Caulfield Marsh. The oldest mine workings, known as the Beaver Hill No. 1 or Graham Mine, consisted of a water level tunnel and two lower-level gangways. These gangways extended southwest to the edge of section 17. The water-level tunnel went completely through the ridge southwest to the ravine opposite the Klondike Mine.[17]

Beaver Hill Number 2 consisted of water-level gangways that ran northeast to the edge of section 17. The three upper gangways extended through the southeast corner of section 8. Beaver Hill No. 3 lay below the other two mines. The workings in

No. 3 consisted of gangways running northeast and southwest from a 3,013 foot slope. When the owners closed the mine there were three north gangways. The lowest gangway extended to the section line, and two south gangways reached to within one thousand feet of the southern edge of section 17. In 1944 Oregon State Department of Geology and Mineral Industries surveyed a considerable block of coal below the old workings of No. l and No. 2 mines.

The Beaver Hill bed sloped north 40 degrees, 60 degrees east, and in the northeast corner of the section it turned around still farther to north 75 degrees east. The dip at the surface varied from 40 degrees to 45 degrees. It gradually flattened downward along the main gangway to 26 degrees in the lowest workings of the mine, some 1,400 feet below sea level. The roof above the coal consisted of a fairly hard sandy clay that showed some disposition to crumble. The floor of the upper mines was a hard tenacious sandstone but in the lower workings this changed to shale overlain by soft clay. This floor "swelled" with a tendency to push up into the tunnel from the bottom.[18]

Geologist J. S. Diller made an extensive survey on the coals in Coos County in 1899. His report showed that L. A. Whereat, a civil engineer who helped bring in the Southern Pacific lines to the area in 1916, was working for Graham at the Beaver Hill mines. Not a mining engineer, Whereat, however, managed the mines for a number of years. L. A. Whereat was the grandfather of Don Whereat of Kentuck Slough. Whereat hired George W. Evans, a mining engineer from Seattle, as a consultant. Evans introduced some new mining methods and safety regulations.[19]

Diller also noted in his report the amounts of coal and the various beds in the area, as well as the serious faulting of the earth nearby. The earth faulted and slipped as much as twenty feet at places in the Beaver Hill Mine.[20] The faulting, the pumping of water, the ventilation costs, and the pushing up of the floor at the great depths made the largest coal mine in Coos County a very expensive operation. Yet, in spite of all this, the Southern Pacific Company operated the mine until 1923.[21]

The tipple (left) at Beaver Hill sorted coal by size as it came from the shafts. The hotel, store, and workers' housing stood on pilings and on the nearby hillside at the head of Caulfield Marsh. (Oregon Historical Society Neg. No. 38865)

Beaver Hill averaged 200 tons per day throughout its life. Miners produced 500 tons a day for twenty days during March, 1905, and one day they reached 600 tons.[22]

June 29, 1911, the *Marshfield Sun* reported, that "The gasoline sloop, *Owl*, sailed Tuesday for San Francisco, loaded with 50 tons of coal from the Beaver Hill bunkers," indicating that production had continued well into the twentieth century.[23]

Beaver Hill was not without its tragedies. On Wednesday, November 1, 1916 at 8:30 P. M., a terrific explosion occurred, where twelve men were working at the 2,000 foot level. The explosion killed Martin Bomar, Cyrus Ferreri and John Lowe. Other men suffered injuries.[24]

Bomar was forty-nine years of age and had worked at the mine for several years. He left a wife and six children who all lived at Beaver Hill. His mother, Carrie C. Bomar, lived with a daughter, Mrs. J. S. Moore, in North Bend. Bomar was born in

Iowa. When yet a young man he met and married Helen Rhode in Iowa. At age twelve, Helen had emigrated with her parents from Germany. In 1992 their daughter, Helen (Bomar) Gorbutt, told of living in Beaver Hill when she was a young girl. Helen Gorbutt was the mother of Merle (Gorbutt) Ellingson and Marilyn (Gorbutt) Stevens.[25]

Cyrus Ferreri had a ranch on South Inlet where his family continued to live for a few years after his death. Ferreri was forty years old at the time of the accident.[26]

John Lowe, known as "Scotty" to his friends, was forty-eight-years-old and had spent a number of years working at the mine. He left a wife and two daughters in Oakland, California.[27]

Thomas Harrison received minor injuries and Thomas McKelvie had face and back injuries but not seriously. Ferreri and Lowe were driving a new tunnel from the lower level and had gone back into the gangway after eating lunch. Bomar, Harrison, and McKelvie were still eating. Suddenly the explosion blew them down and threw them about. Blinded, they staggered toward the shaft, and, while on the way, they heard Bomar cry out that he could go no farther. The deadly after-damp that inevitably follows a methane explosion overcame him. Harrison and McKelvie, however, made their way to the shaft and sent the alarm.[28]

The mine whistle immediately alerted the whole town and the workers began the rescue. They cleared the mine of the poisonous gases and brought out the bodies. The effort took several hours. Meantime, Whereat summoned Dr. Everett Mingus of Marshfield to the mine. Earlier, Whereat had instituted a rule that employees must test the gangways every half hour for gas. Always, after the miners had been away for any time at all, they tested for gas. The miners knew that over a 1,000 feet below sea level gas became a serious problem. Gas quite often broke through and filled an entire area.[29]

In 1913 the state established the Oregon State Industrial Accident Commission. Five men killed in a logging train accident on Bear Creek near Bandon that year provided some of the motivation for the commission. At first the commission gave the widows of the slain men $30.00 monthly throughout their lifetimes, or until they remarried. They awarded each child $10 per month to age 16.[30]

Government authorities sent J. J. Corey of the U. S. Bureau of Mines from his station in Seattle to inspect the mine and check safety conditions. His inspection also included observations regarding any possible damage to the mine which might cause a problem in the future.[31] Corey later became superintendent of the mine.[32]

After the accident Thomas McKelvie moved his family to Wyoming and worked in the mines there. In a couple of years he returned to Beaver Hill and remained until the mines closed in 1923. Thomas had six children: Mary, Rose, Alex, Margaret, Josephine, and Thomas.

October 14, 1921, twenty-three men made up the work crew. Suddenly, and without any warning, a tremendous explosion of methane gas hurled the men about and covered some with debris. Others received severe burns. The explosion occurred about 10 o'clock in the morning. Injured in the explosion were Thomas Hale, foreman of the crew, Thomas McKelvie, Jack Abbot, Leonardo Garagaza, John Pete, Archie Delen, George Levitt, Richard Marsden, Elva Fowle, Joe Menegat, Albert Menegat, Lincoln Yocum, Thomas Kingsley, L. Gonzales, Charles Hill, Carl M. Bills, Julius Fink and Louis Bosco.[33]

Immediately those not too seriously burned gave the alarm and the mine whistle at the adit started a series of sharp whistles announcing an emergency. The whistle alerted the whole town of Beaver Hill. J. J. Corey, superintendent of the mine, arrived at the mine adit about ten minutes after the explosion. He immediately took charge of the rescue work.[34] The night shift ran to the mine to help. Teachers dismissed school and families gathered at the mine entrance. Doctors in Coquille and Marshfield reported and cared for the injured. Drs. Hamilton and Low came from Coquille. Repeatedly people in Coquille telephoned Beaver Hill to determine the extent of the accident. They learned only that bodies were being brought to the surface.[35]

Meantime down in the mine Thomas Kingsley busied himself seeing to the needs of the severely injured. Joe Menegat and Pete Marsden, who were also in the mine, soon joined him. Later, members of the night shift came to the rescue. Joe Menegat received some burns and assisted in the rescue. Joe Menegat's burns were not serious enough to cause him to be hospitalized. Marsden received a scalp wound during the explosion but was not burned. He also was not hospitalized. Fred Denny and Eugene Bosco soon came to the scene bringing additional face masks, safety lamps and a pull motor to use on the ones suffering from the deadly "after-damp." The rescue work was very difficult. They uncovered the rocks and coal from Julius Fink and Liberio Gonzales.[36]

The rescue team checked the air shaft to make sure fresh air was available. Also the men eliminated the gas as rapidly as possible. Several hours elapsed before they evacuated all the injured. Except those gassed, none of the injured was unconscious when rescued.

Drs. Hamilton and Low took charge of the first six men brought to the surface and quickly spirited them to the Coquille hospital. They were John Pete, Archie Delen, George Levitt, Richard Marsden, Elva Fowle, and Carl Bills. Thomas Hale, Thomas McKelvie, Leonardo Garagaza, Albert Menegat, Lincoln Yocom, Thomas Kingsley, Liberio Gonzales, Charles Hill, Louis Bosco, Elva Fowle, and Julius Fink went to the Mercy Hospital in North Bend.

Thomas Hale, the foreman, was sixty-two-years-old and was one of the more seriously injured. Albert Menegat was thirty-years of age and had a wife and four children at Beaver Hill. Lincoln Yocum was severely burned. He was twenty-two-years-old and lived with his sister in Beaver Hill. Elva Fowle was thirty-two-years of age and was very badly burned. Thomas Kingsley was reported as seriously burned.[37] Later it was found that Kingsley was injured but was not burned.[38]

Louis Bosco and C. M. Bills were gassed; however, Bills was in good enough condition to be released from the hospital on October 17. He and his wife were parents of a new baby.[39] Bosco was a single man twenty-three-years old. His brother, Eugene Bosco, helped in the rescue.

Liberio Gonzalas was twenty-one-years of age and lived

with his sister. Gonzalas suffered a broken arm besides the burns. The rescue team dug him out from under rocks and coal. Leonardo Garagaza also suffered serious burns. He was twenty-seven-years-old and married but had no children.[40]

Thomas McKelvie, who had not been injured in the 1916 explosion, was not so fortunate this time. He was also severely burned. His thirteen-year-old son, Alex, watched at the mine entrance. Alex, in his mid-80's, remembered quite vividly standing at the mine entrance the day of the 1921 explosion. He watched, like many others, while the workers brought out the bodies of their comrades and the injured.[41] Alex and his mother visited the Mercy Hospital while his father was hospitalized.[42]

Liberio Gonzalas died October 16 at Mercy Hospital in North Bend. Gonzalas was a native of Mexico and had been making his home with his sister, Yizarria Gonzalas, in Beaver Hill. He and his sister had been in Beaver Hill about six months. Authorities shipped his body to his father in Mexico.[43]

Within a few days men in the Coquille hospital, except George Levitt, showed improvement. Some feared Levitt would succumb from the injuries. He was severely burned. Members of the State Industrial Accident Commission came to the area and checked the extent of the injuries. November 1, Mercy Hospital reported that all of the men there were improving except Thomas Kingsley. Kingsley was suffering from a serious infection.[44]

Dr. Mingus made an appeal to the America Legion for volunteers to help in the nursing. The sudden influx of so many injured men taxed the hospital staff beyond its ability to do the necessary work. Dr. Mingus also installed specially built tanks filled with a soda solution in which the men placed their arms to help in the healing. Leonardo Garagaza was discharged from the hospital on November 8.[45]

Albert Menegat died at 11:30 A. M. on October 21, a week after the explosion. Menegat was born in Italy and had lived for a while in Kentucky and Arkansas. He worked in the mines in those states before coming to Oregon. Albert's wife was Egidia (Sacate) Menegat and their children were John, Mary, Eleanora, and Rolando Menegat. Later, Egedia married Eugene Bosco and their son was named Joseph Eugene Bosco.

John Menegat lived for a short time with his uncle in Lakeside. When his mother remarried he moved back home. Bosco was working and living in Conlogue's logging camp. The Conlogue camp was just over the hill southwest of Beaver Hill. John and some of the other young boys explored the abandoned Beaver Hill buildings in 1924.[46]

November 28, 1921, the Mercy Hospital reported that George Levitt had been admitted there. The Coquille Hospital had released Levitt, but his condition worsened. After a few days, however, Levitt recovered. The other victims, except Thomas Kingsley, also recovered. Mercy Hospital reported that Kingsley had become partially paralyzed due to a hemorrhage in the spinal cord. He succumbed the following day and was buried December 4, 1921, in Sunset Cemetery near his co-worker Albert Menegat.[47] His widow, Mary, later married August "Gus" Matson. They had a son, Philip A. Matson. Besides his widow, Mary, Thomas Kingsley left two children, Thomas, and Emily, both born in Scotland.

November 7, 1921, J. E. Z. Fowler recorded the official Coos County school census for Beaver Hill School District #69. Although the area was known as Beaver Hill, they named the post office Preuss after the owner of the local store. Many said a thousand people lived there in the early twenties; however, the school census and the number of houses indicated considerably fewer people. The school census included all children from age four to twenty and was comprised of thirty-two boys and forty-eight girls. Nine of the children in the census lived at Leneve and attended school there. Leneve was about five miles southwest of Beaver Hill on the Coquille River.

School District 69 census for 1921 read as follows:

Abbott: *Laura, 11, Eva 10, Lorance 8, Minnie 7, father, John Abbott;* Barnett: *James 5, father, James Barnett;* Conlogue: *Alice 6, Irene 4, father, D. J. Conlogue;* Coney: *Edward 20, Edith 5, Florence 19, father, J. J. Coney;* Demetrovick: *Mike 10, Annie 6, Angelina 7, August 5, father, Augustine Dometrovick;* Eckberg: *Lillian 4, mother, Donna Eckberg;* Echloms: *Oke 6, Margaret 5, father, Sig Echloms;* English: *Lewis 20, father, H. M. English;*

Engbloom: Violet 18, father, W. D. Engbloom; Erickson: *Reina l0, parent, Frank Granfors;* Frullany: *Thomas 10, father Joe Frullany;* Fry: *Clarice 14, Florence 10, Howard 5, Muriel 12, father, Clarence Fry;* Gransfors: *Dianna 4, father, Frank Granfors;* Griffin: *John 20, father, Paul Griffin;* Grove: *Jack 5, Arnold 7, father, Carl Grove;* Hicks: *Luella 8, Olive 6, Ruth 4, father, Jack Hicks;* Houser: *Hilda 4, father, Tines Houser;* Hurme: *Mary 16, father, Frank Hurme;* Keyes: *Lillian 12, father, J. M. Keyes;* Kingsley: *Emily 8, Thomas 6, father, Thomas Kingsley;* Kronsteiner: *Carl 6, Joseph 9, Margaret 12, father, J. Kronsteiner;* Lizaraga: *Marguerhitta 7, mother, Inez Lizarraga;* Luna: *Christine 17, father, Ignacio Luna;* McKelvie: *Mary 15, Alexander 13, Josephine 6, Margaret 8, Rose 12, Thomas 4, father, Thomas McKelvie;* McKenna: *William 5, Cyril 6, father, W. P. McKenna;* Menegat: *John 7, Mary 13, Nora 4, father, Albert Menegat;* Morey: *Erma 19, parent, Elizabeth Royer;* Nelson: *Gordon 19, father, T. Nelson;* Pavlinac: *Allina 4, Joseph 11, father, R. Pavlinac;* Perkins: *Kenneth 4, mother, Mildred R. Perkins;* Pete: *Vivian 18, parent J. M. Keyes;* Rodgers: *Daniel 5, Milton 7, Rachael 4, father, Daniel Rodgers;* Rogers: *Geraldine 7, Maxine 9, father H. B. Rogers;* Rogers: *John 6, Velma 7, father, Fred Rogers;* Rollins: *Ralph 18, parent, Linnie C. Fowle;* Simmons: *Margery 5, father, Fred Simmons;* Sodden: *Ruth 18, father Sodden;* Teal: *Henry 5, Nellie 10, father, Harry Teal;* Thompson: *Frank 7, Mary 6, Nina 9, father, Ed Thompson;* Webster: *Dayle 6, Faye 8, Lila 11, Olga 4, father, Guy Webster;* West: *David 8, mother, Meda Thompson;* Crnkovich: *Annie 5, parent R. Crnkovich.*[48]

The Beaver Hill Mines and its activities had been a vital part of Coos County for over twenty-five years. The hamlet of Beaver Hill in the early part of the twentieth century had taken on an air of permanency. The mine produced more coal than any of the other mines in the county. The community grew considerably during the last ten years of its existence. Beaver Hill operated a four room school house under the leadership of County School Superintendent Charles A. Mulkey. The town boasted a

In 1916 the Beaver Hill School had four teachers and good enrollment. Several students gathered around Charles A. Mulkey (left), Coos County School Superintendent. Mine explosions in 1916 and 1921 sealed the fate of the mine and community. (McKelvie Collection)

The Beaver Hill Saloon survived the Volstead Act of 1919 but no longer served beer. The building continued as a "dry" social club in the early years of Prohibition. (Slattery Collection)

Beaver Hill possessed 152 residents in 1910. The company town overlooked Caufield Slough. This thriving Coos County coal mining town vanished in the 1920s with the closing of the Beaver Hill Mine portals and the falling off of the coal markets. (McKelvie Collection)

hotel, a barbershop, a store with a post office, a pool hall that formerly had been a saloon, and had many white-painted houses. Situated on a low lying ridge in the coast mountains, the town consisted of houses built on six levels connected with stairways between levels. An empty box-car served as a jail for the unruly and miners who drank too much moonshine, the local beverage during prohibition days. Some families kept a cow and raised gardens on the bottom beside Caulfield Slough.[49] Within a few years after the mine closed in 1923, not a vestige of the once thriving community remained. Beaver Hill became another of Coos County's disappearing towns. The Southern Pacific Company dismantled most of the buildings and took some of the lumber to Roseburg.[50] Later Bill Furbrasch, Pud Nosler, Josh Ruble, Roy Pettingill, and others tore down the remaining buildings and used the lumber to build several houses in other localities.[51]

KLONDIKE MINE

The Klondike Mine opened southwest of Beaver Hill in section 18. The adit was a few feet above sea level about a mile north of the Coquille River. The mine first opened during 1898 and 1899. Reports indicated the mine tunnel descended 1,000 feet to the southwest. The operators failed to keep reliable records of production. After a period of inactivity, miners worked it briefly again in 1932. At that time the coal caught fire so the miners pulled the pillars. Coal left in the rooms to support the roof was called pillars. In an unpublished report from the files of the Southern Pacific Company, dated October, 1921, G. J. Sielaff stated that the mine produced five tons a day. Sielaff reported that the tunnel reached into the mountain 400 feet. Two gangways provided the coal.[52]

Inhabitants of the small hamlet at the Klondike Mine called the place Beaver City. During World War I the U. S. Army's Spruce Division built a branch line from the Southern Pacific main line into the Klondike. The army moved in loggers, called the Spruce Division, to log spruce for airplane construction. When the war was over, the army left and abandoned most of the machinery and the railroad tracks. The Conlogue brothers logged near there,

J.S. Kaufman, Marshfield banker, handled payrolls of the local mining companies prior to World War I. (Stadden Collection)

so they used the railroad to haul logs to the river. Conlogues worked the mine in a small way to obtain coal for their locomotive and the cook houses and bunkhouses at their logging camp.[53]

When Edna (Lemon) Gilfillan, wife of John Gilfillan, was a little girl she went with her father to dig coal from the mine. They used the coal in their house at Conlogue's logging camp. Her father, Claude Lemon, was working for Conlogue brothers.[54]

Closing of the Beaver Hill Mine and the intermittent working of a few mines like the Southport, the Overland, the McLain and Alpine showed the decline of the coal industry in southwestern Oregon. The coal fields at Beaver Hill, however, had demonstrated the region's fuel potential and tested the ingenuity of miners and the capital of investors. No other Coos county mine plunged to more than 1,000 feet below sea level. No other mine took so heavy a toll in human life.

ENDNOTES

1. Lynne Bowen, *Boss Whistle*, pp. 15-17.
2. Ibid., pp. 15-20.
3. Ibid.
4. Mitzi Asai Loftis, *Made in Japan and Settled in Oregon*, p. 128.
5. James G. Masters, enumerator, Twelfth Census of the United States, Coaledo Precinct, Beaver Hill Village, Oregon. Schedule No. 1., 1900; Garfield C. Thompson, enumerator, the Thirteenth Census of the United States, Beaver Hill, Oregon 1910.
6. *Coos Bay Harbor*, November 1, 1915.
7. Bowen, op. cit., p. 20.
8. Ibid., p. 50.
9. Karl Alpine, Interview with Dow Beckham, March 20, 1993.
10. Bowen, op. cit., pp. 50-51.
11. Ibid., p. 49.
12. Ibid., p. 50.
13. Emil R. Peterson and Alfred Powers, *A Century of Coos and Curry*, p. 399.
14. Boyd Stone, Beaver Hill Mines, *Coquille Sentinel*, October 14, 1992.
15. Peterson and Powers, op. cit., p. 399.
16. Coos Bay News, February 5, 12, 16, and 22, 1896.
17. Peterson and Powers, op. cit., p. 400.
18. John Eliot Allen and Ewart M. Baldwin, *Geology and Coal Resources of the Coos Bay Quadrangle, Oregon*, pp. 92-93.
19. Peterson and Powers, op. cit., p. 400.
20. *Marshfield Sun*, June 29, 1911.
21. Peterson and Powers, op. cit., p. 400.
22. *Coos Bay Times*, November 1, 1916.
23. *Marshfield Sun*, June 29, 1911.
24. *Coos Bay Times*, November 1, 1916.
25. Helen Gorbutt, Interview with Dow Beckham, March 23, 1992.
26. *Coos Bay Times*, November 1, 1916.
27. Ibid.
28. Ibid.
29. Ibid.
30. Ibid., November 3, 1916.
31. Ibid.
32. Ibid., October 15, 1921.
33. *Coos Bay Times*, October 15, 1921.
34. Ibid.
35. *Coquille Sentinel*, October 14, 1921.
36. *Coos Bay Times*, October 15, 1921.
37. Ibid.
38. Ibid.

39. Ibid.
40. Ibid.
41. Alex McKelvie, Interview with Dow Beckham, March 14, 1992.
42. Ibid., March 25, 1992.
43. *Coos Bay Times*, October 15, 1921.
44. Ibid.
45. Ibid., November 9, 1921.
46. John Menegat, Interview with Dow Beckham, August 3l, 1992.
47. Sunset Cemetery Records.
48. Coos County Census Records, Educational Services District.
49. Boyd Stone, "Beaver Hill Mines," *Coquille Sentinel*, February 14, 1992.
50. Ibid.
51. Boyd Stone, *Myrtle Point Herald*, February 14, 1992.
52. George Welch, Coos County Coal Mines, MS, p. 16.
53. Ibid., p. 16.
54. Edna (Lemon) Gilfillan, Interview with Dow Beckham, October 27, 1992.

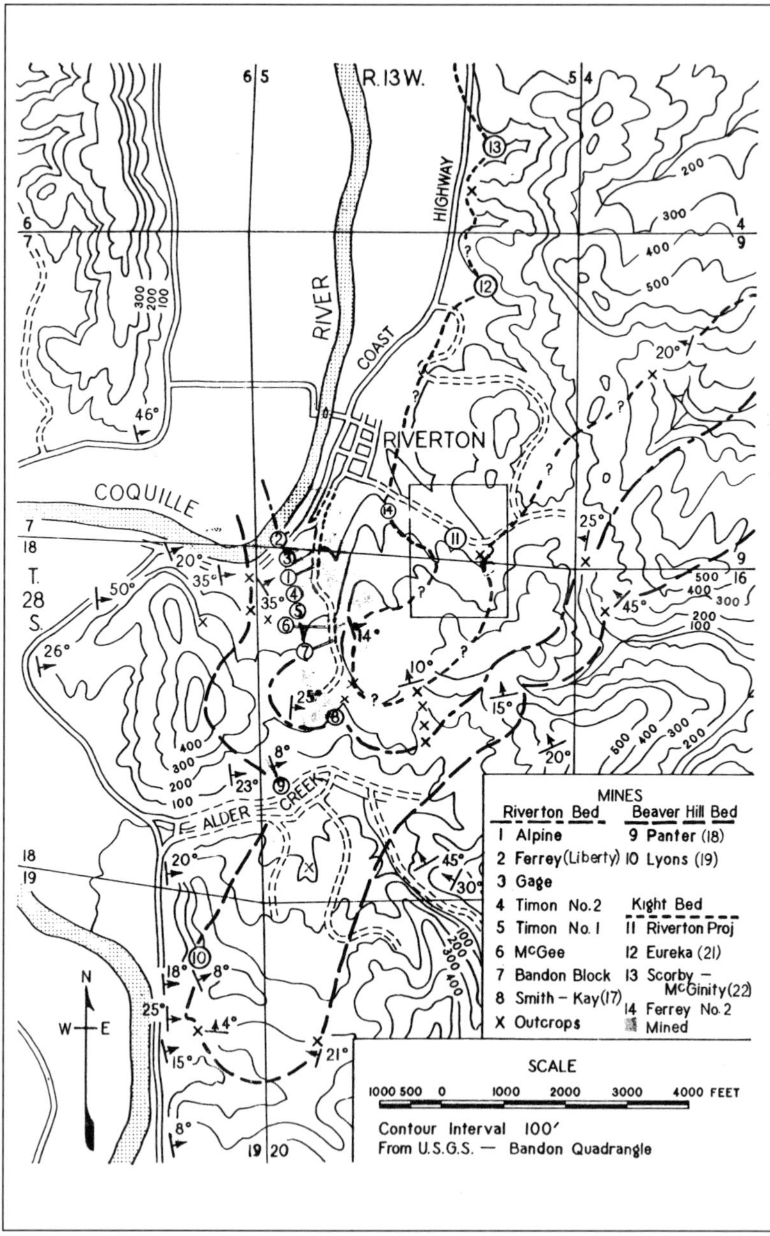

The coal deposits at Riverton encouraged developers to sink a number of shafts along the banks of the Coquille River. These mines tapped outcrops and deeper deposits. The investments gave rise to the small town of Riverton. (Allen and Baldwin 1944:100)

SEVEN

COQUILLE RIVER MINES, OTHER COALS & INDUSTRY CHANGES

While the coal fields in the Coos Bay watershed drew the earliest investment and development, those along the Coquille River in southern Coos County also proved productive. Coal from these deposits served both a local market as well as export across the Coquille River bar at Bandon to distant ports of call.

MINES NEAR COQUILLE

The Peart Mines operated in the years of 1908-09. The Old Peart Mine, located just to the north of Coquille, produced ten to twenty tons of coal per day. The mine was in the SW1/4, NW1/4, section 36.[1] The New Peart Mine also opened in that same quarter section. The McClain Mine was located northwest of Coquille in the SE1/4 section 35, T27S, R13W. In 1929 developers spent an estimated $20,000 to develop the McClain Mine. It opened the following year with twenty workers who produced one hundred tons of coal per day.[2]

J. S. Diller also reported coal out-croppings on Fat Elk Creek a few miles southwest of Coquille.[3] Downstream a few miles from Coquille, near the junction of Chrome Road with the North Bank Road, were Fahy-Muir coals. In 1911 geologists J. S. Diller and M. A. Pishel checked these out-crops and noted the quality. The out-crops were in the N1/2, section 30, T27S, R13W. In 1943 Allen and Baldwin checked and measured six coal out-crops in that same locality with conclusions comparable to those of Diller and Pischel.[4]

RIVERTON BASIN COALS

The Riverton district produced coal for over fifty years beginning with the Timon Mine in 1895. Between 1900 and 1914, miners also opened the Eureka Mine near Riverton. It was north of Riverton about a quarter of a mile and east of Highway 42 in the NE1/4, section 8, T28S, R13W. The Riverton Basin deposits produced most of the coal along the Coquille River. Geologists placed the Riverton coals as a part of the Beaver Hill bed. At Riverton several mines successfully provided coal for local consumption and shipment to the California markets.[5]

Joseph Ferrey opened the Liberty Mine at Riverton, near the bank of the Coquille River, in 1897. The slope of the Liberty Mine went down 1,300 feet on a twenty degree dip with four gangways. The mine produced between fifty and seventy-five tons a day until 1902. Water flooded out the miners that year.

The Arcata *lay low in the Coquille River upon taking on a cargo of coal at the Kay Mine at Riverton. The bunker with open walls stood on the bank at the terminus of the trestle from the mine. (Bandon Historical Society)*

Later, the owners pumped out the gangways, opened them and operated until 1914. The Gage Mine was between the Liberty and the Alpine on the line between sections 17 and 19. Miners sank a shaft 300 feet at a steep angle to find the coal. They extended a gangway north to the Liberty workings and another gangway south 200 feet.[6]

In the years 1898 to 1900, miners operated the Bandon Block Mine near Riverton. They opened the tunnel on the north side of the ridge and dug all the way through. The mine contained some rooms 150 feet in length. In cavities of this size, miners left several pillars of coal, carefully spaced, to support the roof. William H. Kay purchased this property in 1941. In 1942 he leased it to J. D. Kay and E. M. Smith who produced from five to fifteen tons of coal a day during the next two years.[7]

In 1907 John R. McGee took over a mine in Riverton and operated it until 1921. McGee was born in Missouri, September 17, 1869. The Klondike gold fields beckoned to many young men and McGee and a friend felt the lure and headed west. They got as far as southwestern Oregon on their way. They found work in the coal mines of Libby. A short time later a glitter of another kind caught the eye of John McGee. He fell in love with Catherine Joffey. They were married in 1899 and thus ended the trip to Alaska. Mining, however, became McGee's occupation. He worked in the Libby mines and later for James Flanagan at the Southport Mine and at the Beaver Hill Mine. He then moved to Riverton to run his own mine. While at Beaver Hill, young McGee played in the local band. His daughter, Esther (McGee) Phillips, was born near the mouth of Beaver Slough, April 1, 1904.[8]

During the early years, Riverton miners shipped coal to the San Francisco Bay market. Sailing ships came up the Coquille River as late at 1922 for coal and lumber products. The *Arcata*, one of the early ships to haul coal from Coos Bay, also took loads out of the Coquille River from the Riverton mines.[9]

THE ALPINE COAL COMPANY

W. T. Alpine moved his family to Riverton in 1921 where he set up the Alpine Coal Company. After contracting with other

The McGee Bunker (right) and the Alpine Bunker (left) confirmed that in 1915 coal played a major role in the economy of Riverton. The mines produced thousands of tons of coal over 50 years of operation.

mines for a number of years, he opened his own mine in the same vein that McGee mined. The Alpines operated one of the last mines in the Riverton district. The adit to the Alpine Mine opened about one hundred yards east of the highway on the southern edge of Riverton in NW1/4, section 17, T28S, R13W. W. T. Alpine operated the mine from 1930 to 1941 during which time he produced 100,000 tons. Part of the time, the miners used an unusual technique for getting out the coal. They picked out the seven inch center of coal that was contained in "bone" bands of unsuitable material. Once this was done they dug the remainder of the seam without blasting. In 1938 the mine employed ten underground workers who averaged twenty tons a day.[10] Alpine operated the mine under a lease from the Riverton Coal and Development Company. The area covered about 230 acres in sections 17 and 18, T28S, R13W. During most of those years men mined in rooms forty feet wide and supported the ceiling with pillars of unmined coal, periodically. They also used a lot of stulls for holding up the roof. Diggers loaded the cars at the face and used the weight of the loaded car to pull the next

Above. *A precarious trestle transported coal cars from the Alpine Mine to a bunker and spanned what in the 1930s became Highway 42 at Riverton.* (Coquille Herald Files)

Right. *Karl Alpine, 1992, worked in the Alpine Mine at Riverton during the dark days of the Great Depression. By 1938 the mine averaged a production of about 20 tons per day. (Beckham Collection)*

empty car up the inclined floor with a cable through a block (pulley). A small horse pulled two loaded cars to the bottom of the incline. The hoist lifted one car at a time to the surface. From there they pushed them over a tramway to the bunker near the river.[11]

Alpine's son, Karl Alpine, started working in the mines when he was a small boy attending school at Riverton. Karl worked many a weekend in the mines, not seeing daylight from Friday until Monday morning. Karl and the miners went into the mine before daylight on Saturday morning and came out after dark. They did not work on Sundays. However, occasionally they worked the entire weekend. When the Alpines first arrived, they joined a crew of thirty miners digging coal for the San Francisco market. In the later years the Alpines continued working with a limited crew and marketed their coal locally. They sold to the County Courthouse, Riverton School and a few others. Karl Alpine recalled:

Mining was nothing but hard work. The vein was so low that the miners had to work stooped over, or on their knees, or their bellies, or on their backs. They used 'room and pillar' method of mining. They left pillars of coal intact to hold up the roof until a sizable area was dug out before extending the gangway to set up another room. Later, after a series of rooms along a gangway were mined, the miners sometimes 'pulled' the pillars. If the gangway was no longer needed and conditions permitted, they 'pulled the pillars' for the coal which they held.[12]

"Pulling the pillars" was a dangerous maneuver. The miners started at the back and dug out the pillar. The ceiling began to weaken and, sometimes before they pulled the last pillar, the roof fell. The miner quickly scurried to safety.

After the ships quit coming to the mine, the company barged one hundred pounds sacks of coal to Bandon. In Bandon, they loaded the coal on the steamer *Elizabeth*, a small coastwise steamer hauling goods south to the San Francisco markets. Alpine barged other shipments upriver to Coquille and scooped the coal on Southern Pacific gondolas bound for Eugene or Portland. A few years before the mine closed, a customer from Smith River in northern California hauled cream to the Coquille area. For a return load, he took coal to Crescent City. During the Depression, local people drove to the plant and picked up four or five sacks of coal for home use. Alpine charged fifteen cents a sack for nut coal, which was small pieces, and thirty-five cents for lump coal, the larger pieces.

During the Depression Karl Alpine was the delivery boy:

I remember all the sacks of coal I carried to Tommy Wing's restaurant in Coquille. There was only one access to the place and that was the front door. The Coquille Laundry also bought coal from me. We just had a pickup. I made as

The sternwheeler Telegraph *fired her boilers with coal from the Alpine Mine at Riverton. W. T Alpine operated the mine from 1930 to 1941 and constructed the buildings near the shafts on the riverbank.* (Coquille Herald Files)

> *many as seventeen trips a day from Riverton to Coquille. Once, my father went to San Francisco to check the coal market. When his ship, the* Breakwater, *crossed the Bandon Bar, the current swung the boat against the jetty, but it did not suffer too much damage.*[13]

OTHER RIVERTON MINES

Near Riverton, Frank Rouse opened up a small vein about thirty inches thick and built a large bunker on the river. He spent a lot of his money on this venture that failed.[14]

Two sailing schooners, the *Free Trade* and the *Ruby*, hauled coal from the Riverton mines. Steam tugs, the *Katy Cook* and the *Tryumph*, towed these sailing vessels up the Coquille to the coal mines. Sometimes, after loading the holds with coal, the ships completed their shipment with a deck load of lumber. The Price sawmill at Riverton sold products in California also.[15] Two other ships, the *Arcata* and the *Anvil*, hauled coal from the

Above. *The* Anvil, *a steam schooner with sail assist, took on a cargo of coal at Riverton on the Coquille River early in th twentieth century. (Bandon Historical Society)*

Left. *Leonard Gibbs, age 87 in 1992, and his wife, Matilde, ran the last operating coal mine unitl 1955 in Coos County. They sold coal to heat public buildings and homes. (Beckham Collection)*

Coquille River mines.[16] A number of other ships also called on the Coquille River ports for lumber and coal.

Pat McGourty and his son-in-law, Jack Scorby, opened up a small mine on the same vein as the Eureka Mine near Riverton and worked it about ten years. Another venture profited the land owners, the Joseph Donaldson family, who had a ranch across the river from Riverton. The Donaldsons were said to have collected $50 per month for the mineral rights for many years. Under the lease miners dug 1500 feet of tunnel but found no coal.[17]

At one time Riverton was one of the booming coal mine towns of Coos County. In its heyday, it supported seven saloons, a dance hall, and other forms of entertainment for the miners and the loggers who dug the coal and logged the big timber that grew on the hills close around the little town. For a number of years, Riverton had a reputation for being a wild town.[18]

Other mines near Riverton included the Urquhart, Kight, Panter, Gier, Rouse, and Sell developments. The Lyons Mine was on Harlocker Hill in the SE1/4, section 23, T28S, R13W.[19]

After Jack Timon sold out his interests at Riverton, he opened up a mine on the Lampa Coal Company property. The adit opened just below the old Lampa post-office in the NE1/4, section 36, and the SE1/4, section 25, T28S, R14W. The Lampa Creek Mines were approximately five miles downstream from Riverton. Through the years at Lampa Creek, miners prospected seven beds of coal within a stratigraphic distance of 400 feet. Matt Laws operated there for a while, but he gave up and moved away in 1914.[20]

Later, Leonard Gibbs operated the Lampa Coal Company and worked bed number four of its property. An investor named McGurty drove a tunnel 160 feet into the mountain before he reached the coal. He discontinued his work and Gibbs mined there intermittently before moving to South Slough in 1942. Gibbs worked the bed and mined along the strike of the coal until it faulted a short distance from the portal. In 1941, M. R. Geer, for the U. S. Bureau of mines, listed Gibbs's production at 665 tons.[21] Although the geologists Allen and Baldwin numbered only seven beds, Gibbs, however, said that eleven veins of good coal showed at Lampa Creek. Gibbs also said that the area was badly faulted.[22]

Other coals in this area included the Woomer and Donaldson prospects. Also Allen and Baldwin, geologists mapping the area in 1943-44, found coal beds on Seven Mile Creek. They reported one vein seventeen feet thick, there.[23]

BANDON COALS

In early 1941 L. S. Klineman prospected on a small creek that runs through the city of Bandon. He found several outcroppings and encountered coal in test pits. Geologists, Allen and Baldwin, in their 1943-44 report, noted the coal was good quality. They made only a limited survey of the veins.[24]

Allen and Baldwin reported on out-croppings of two thin beds of coal and coal-like material high on the sea cliff about half a mile north of Fivemile Creek in section 32, T28S, R14W. The beds sloped north and west 50 degrees and slanted at 42 degrees. This coal comprised part of the lower coal of Sevenmile group.[25]

In 1943 the Department and Mineral Industries of Oregon traced the course of the coal over 5,000 feet by means of open cuts, short tunnels, and hand auger holes. Allen and Baldwin, doing this work, found that the Beaver Hill coal bed cropped out in the S1/2, section 2, T27S, R14W. In 1932 the McKenna Mine portal lay 1,300 feet west and 800 feet north of the southeast corner of the section. The geologists mapped an area of 150 acres along the southern boundary of the section. They surveyed on a scale of one hundred feet to the inch and a contour at intervals of ten feet.[26]

Leonard Gibbs Mine. Leonard Gibbs and his wife, Matilde, leased their South Slough property from the owners, H. G. Kern, Mrs. Otto Schetter, Groth, and J. S. and M. C. Getty. They established the last active mine in Coos County and continued from the early 1940s to 1955. That year, their last good customer, the Riverton Schools, closed and joined in consolidation with the Coquille district. Gibbs built a truck road from the Charleston-Bandon highway down into Winchester Creek, a tributary of South Slough.[27]

The vein had an excellent roof of good gray sandstone. In three benches, they found a vein about seven feet thick. The vein contained ten inch partings and impurities in the middle of the seam. On the top part, they had about the same amount of very poor grade coal. Fortunately, the vein was level. At first, they had an old International flatbed truck. Matilde shoveled all of the coal at the delivery points. After Matilde got a dump truck, her work was a little easier; however, at some delivery points she still had to shovel it out of the truck. Gibbs delivered coal to

In the early 1900s coal powered lumber schooners and the sawmill at Bandon. The Riverton mines provided the energy for local industry.

a number of public building such as the Sumner School, the Bridge School, and Riverton School. She also delivered to business buildings and private homes.[28]

When the Southport Coal Company closed down, all the machinery went to Coos County for taxes. The county sold this equipment to Leonard and Matilde Gibbs who transferred it to their South Slough operations. The equipment included power-driven coal cutters and a mechanical loader. They closed their mine in 1956.[29]

McKenna Mine. The McKenna or Schetter Mine workings lay approximately on the axis of a north-plunging syncline. Fred Schetter homesteaded this tract in the 1850s. He discovered coal and the family retained the mineral rights for over a century. The portal of the mine opened on the west side of South Slough at an elevation of about fifty feet. Albino "Blackie" Salaszar, who did most of the mining for McKenna, extended the tunnel into the ridge 500 feet in a southwesterly direction about 70 degrees south of west. The coal dipped from 17 to 25 degrees north.[30] William McKenna operated his mine for a short time in the 1920s.[31] When Gibbs was driving his slope through from the other side of the ridge, he had a few problems keeping the air quality right. When he broke through to the old McKenna tunnel suddenly the west wind whistled through the mine in a blast. The wind brought clean fresh air to the face of the coal.[32]

OTHER OREGON COAL BEDS

Coal exists in a number of areas of the state, such as northeastern Oregon and the northern part of the Coast Range. Outcroppings of coal show near Tenmile Lake. A Camas Valley mine produced coal for a brief time in the early part of the century. Several out-croppings of coal surface in the vicinity of Lookingglass about eight miles west of Roseburg. Small beds of coal show on the north fork of the Umpqua River and Cavitt Creek, a branch of the Calapooya River. The Eden Ridge coals south of Powers, Oregon, comprise about 5,000 acres of land, mostly in the Siskiyou National Forest. These coal beds consist of at least five veins, varying in thickness from four to ten feet. They lie in about 300 feet of gently dipping sandstone. Unlike other minerals, investors in these coal beds are required to lease from the Forest Service.[33]

Joseph Silas Diller of the U. S. Geological Survey also reported on coal in the Rogue River Valley:

> *Of the coal fields in the Southwest Oregon the one in the Rogue River Valley is the most conveniently located with reference to the Southern Pacific Railroad. A number of years ago this field was prospected by the Southern Pacific Company, but production did not follow immediately. Since then there has been a great deal of prospecting, especially in the vicinity of Medford and Ashland. As a result the field is known to be about 80 miles in length and only a few miles in width, dipping eastward beneath the Cascade Range.*[34]

PACIFIC POWER AND LIGHT COMPANY

In 1960 Pacific Power and Light Company core-drilled the Eden Ridge field and estimated that 50 million tons of coal could be mined for its projected plant near there. The company drew

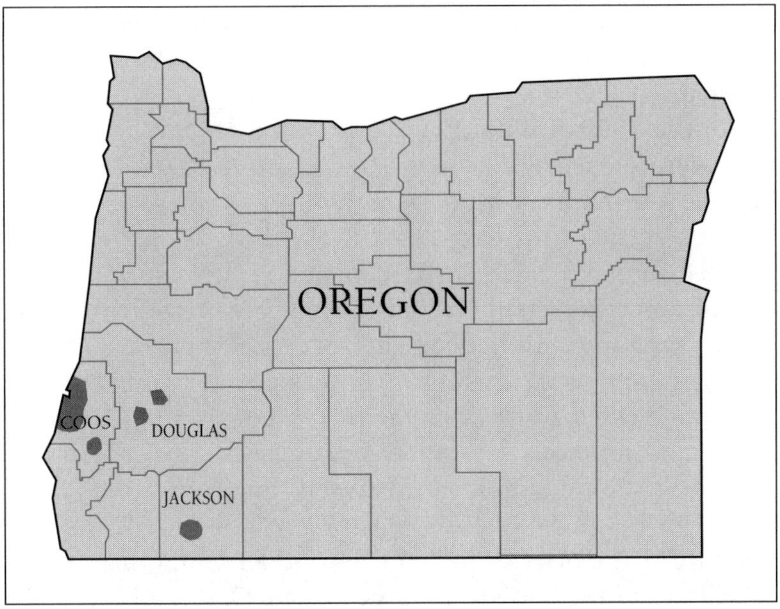

Coal deposits in southwestern Oregon hold the promise for future development as energy resources.

up extensive plans for a power plant. The idea called for a coal-burning steam plant to generate electric power. This was to tie in with a dam on the South Fork of the Coquille River for a cooling system and to produce hydroelectric power in the winter, when the South Fork carries large quantities of water. Calculations in 1960 indicated that these projects would require an investment of approximately $100 million dollars.[35]

Pacific Power, the largest user of coal in the Northwest, operates its steam electric plant at Centralia, Washington with coal mined nearby.[36] In 1910, the Pacific Power & Light Company served just over 7,000 electric customers in towns of Astoria and Pendleton in Oregon, and Yakima and Walla Walla in Washington. Three quarters of a century later, Pacific Power's lines reached more than 665,000 customers in 240 communities in parts of six western states. Its service territory increased to 1,000 miles, from the coastal dunes of Oregon to the high plateaus of Wyoming, and 500 miles from northern Idaho and Montana to northern California. Pacific Power acquired a number of smaller companies and in 1954 merged with Mountain States Power Company. In 1955 it also merged with California Oregon Power Company (COPCO).[37]

In the 1990s Pacific Power generated seventy percent of its power from thirty-two hydroelectric projects and four coal-fired power plants, with a capacity of 4,000 megawatts. The largest electric utility in the Pacific Northwest, Pacific Power, in 1984, founded PacifiCorp. It created PacifiCorp when coal mining and the telephone businesses belonging to Pacific Power grew into full fledged enterprises of their own. Along with Pacific Power's own electric service operations and related businesses, until 1993 PacifiCorp consisted of NERCO, a coal and minerals mining company, and Pacific Telecom, the nation's sixth largest non-Bell telecommunications company.[38] In February of 1993, PacifiCorp announced the sale of NERCO at a price of $470 million to RTZ America, Inc., a subsidiary of the British resource giant RTZ Corp PLC. They operated their Wyoming and Montana coal mines at a profit. However, losses from other mining activities apparently caused the reorganization.[39]

A. J. "Pi" Moore moved to Coos Bay in February, 1959. Pacific Power was testing the coal beds on Eden Ridge. Moore participated in the planning and sampling the coal beds. Moore became district manager in southwestern Oregon for Pacific Power and served in that capacity until he retired in August of 1969. The company bought property in the vicinity of the coal veins and spent huge sums of money in its testing program before finally deciding to abandon the project.[40] During those years Pacific Power built its coal-fired electric power plant at Centralia, Washington.[41]

Tests in the company's Wyoming and Montana mines showed veins of very high-ranking coal, from twenty-five to forty feet thick. Near their power plant at Wyodak, Wyoming, the company found coal beds seventy to ninety feet thick.[42] Southern Oregon coal beds are inferior in rank and size to the Wyoming coals. The comparative differences in these coal veins may have cooled Pacific Power's plans for the Eden Ridge project. Nevertheless, Pacific Power is in a position to some day take the lead in promoting a factory for making products from coal, or take advantage of new developments in coal as liquid fuel.

Pacific Power has a reputation of promoting new undertakings for the purpose of selling energy. In the late 1950s, Pacific Power completed the exploration of the sand dunes north of Coos Bay for fresh water supply. A number of local people knew of the water in the dunes, but Cecil Holland placed a pump in the dunes one summer to prove the presence of an immense supply. It took Pacific Power's capital and engineering expertise, however, to verify a quantity of water necessary for industrial use. During the exploration process, Pacific Power brought an eminent hydrologist from The Netherlands to consult with their engineers. On the basis of Pacific Power's findings, Menasha Corporation built a pulp mill in the dunes. Pacific Power turned the water system over to the Coos Bay-North Bend Water Board. C. P. "Ted" Davenport engineered the water exploration project for Pacific Power and Light Company. Bernard Finnell also worked with Davenport on the project.[43]

PacifiCorp is very much aware that its coal burning power plants waft an estimated 13 million tons of carbon annually into the atmosphere over six northwestern states. In 1994, the company drew plans to plant hundreds of acres of Douglas Fir seedlings in hope of offsetting the damage of these emissions to the ozone layer.[44] Like many other companies it will, no doubt, in the near future look to cleaner burning fuels made from coal such as that developed by Dr. Douglas Keller of Syracuse University.[45]

ENDNOTES

1. John Eliot Allen and Ewart M. Baldwin, *Geology and Coal Resources of the Coos Bay Quadrangle*, 1943-44, p. 111.
2. *Coos Bay Times Supplement*, January 30, 1920, p. 32.
3. Ibid., p. 109.
4. John Eliot Allen and Ewart M. Baldwin, op. cit., pp. 97-98.
5. Ibid., p. 99.
6. Ibid.
7. Ibid., p. 105.
8. Esther (McGee) Phillips, Interview with Dow Beckham, March 3, 1993.
9. I remember seeing sailing ships in the harbor at Bandon in 1922 taking on coal and lumber. I do not recall having seen the *Arcata*, nor do I find any text regarding the *Arcata* hauling coal out of the Coquille River. A photograph on file in the Bandon Historical Society shows the *Arcata* loading coal at Riverton.
10. Allen and Baldwin, op. cit., p. 99.
11. F. W. Libbey, *Progress Report on Coos Bay Coal Field*, 1937-40, p. 7.
12. Karl Alpine, Interview with Dow Beckham, April 15, 1992.
13. Alpine, op. cit., March 13, 1993.
14. George Welch, MS, Coos County Coal Mines, p.11.
15. Ibid., p. 12.
16. Photographs in the Bandon Historical Society with captions.
17. Welch, op. cit., p. 13.
18. Ibid., p. 10.
19. Allen and Baldwin, op. cit., pp. 134-135.
20. Ibid., p.14.
21. Ibid., p.133.
22. Leonard Gibbs, Interview with Dow Beckham, September 15, 1992.
23. Allen and Baldwin, op. cit., pp. 134-135.
24. Ibid., p. 138.
25. Ibid., p. 134.
26. Ibid., p. 111.
27. Leonard Gibbs, op. cit., September 15, 1992.
28. Ibid.
29. Welch, op. cit., p.19. Welch wrote that the machinery reverted to the county when Gibbs closed down his mine, but Gibbs disagreed in our interview.
30. Allen and Baldwin, op. cit., p. 113.
31. William McKenna, Jr., Interview with Dow Beckham, December 18, 1992.
32. Leonard Gibbs, op. cit., April, 1992.
33. J. S. Diller, *Mineral Resources of Southwestern Oregon, U. S. Geological Survey Bulletin, No. 546*, p. 139.

34. Ibid., p. 140.
35. John Svicarovich, *The Pacific Power Story*, p. 35.
36. Ibid.
37. Ibid.
38. Ibid., p. 3.
39. *The Oregonian*, Portland, Oregon, June 3, 1993.
40. A. J. "Pi" Moore, Interview with Dow Beckham, October 7, 1992.
41. John Svicarovich, op. cit., p. 32.
42. Ibid.
43. Ibid., pp. 23-36.
44. *The Oregonian*, Portland, Oregon, March 12, 1993, p. B6.
45. Dr. Douglas Keller, Interview with Dow Beckham, January 26, 1993.

Visionaries have developed several proposals to use the coals of southwestern Oregon. Vern Gorst, a visionary in transportation, in 1914 built his "land and water machine" at North Bend. A pioneer in bus and air transport, Gorst held the first West Coast air mail contract.

EIGHT

LIQUID COAL

Beginning in 1907, the southwestern Oregon coal mine operators experienced a brief period of optimism. In June of that year, the city of North Bend granted a franchise to a new firm organized to sell gas and electricity. Louis J. Simpson, Seymour Bell, and Henry Hewitt, Jr., organized the Coos Bay Gas and Electric Company. They received franchises from North Bend and Marshfield to provide gas and electric power. The plant, built at the south end of Tremont Street in North Bend, mixed local coal with crude oil shipped in from California.[1] The plant cost approximately $125,000. An oil tank holding 350,000 gallons added to the expense. The investors also ordered extra boilers for the Porter sawmill and new dynamos to supply electricity for the new company. The Coos Bay Gas and Electric Company could not develop a market for the gas and electricity as projected and the project failed. In 1918 Mountain States Power Company, in conjunction with C. A. Smith Lumber Company, provided most of the electricity for the area. Simpson's Old Town mill also furnished electric power. In 1917 workers dismantled the coal-gas plant. It had been idle since 1913.[2]

Early in twentieth century, engineers began experimenting with coal and using its chemicals for other products. One such engineer was Rudolph Hillstrom of Coos Bay, who in 1930, revealed that he had been experimenting with coal. Hillstrom hoped to find other uses of its chemical contents. "Rudy," as his friends knew him, was a graduate of Oregon State College Engineering School. One of his achievements was a cranberry-picking machine. Many cranberry growers bought and used the Hillstrom berry picker. The method of flooding the bogs and floating the berries subsequently proved to be a better way.

No doubt Hillstrom obtained some of his ideas about coal from the Germans. Germany began a retort system in the low grade lignite coal after the country lost control of the Saar Basin and its rich layers of anthracite coals. In the 1920s two German chemists, Franz Fischer and Hans Tropsch, developed a process of converting a mixture of hydrogen (H2) and carbon monoxide (CO) gases into hydrocarbons for use as motor fuel. During World War II, Germany used the process extensively when her oil supplies were limited. Over seventy years later chemists continue to refine the Fischer-Tropsch process.[3]

Hillstrom's experiments showed encouraging results. He made the following statement to the *Coos Bay Times* in January, 1930:

> *Contrary to general opinion, the great bulk of natural products from Coos Bay coals are not suitable for motor fuel. This, however, is not necessarily a disadvantage. They can be converted to motor fuel by further processing, but that is not advisable, for the reason that they are worth more in their crude state. It would be foolish, for instance, to convert creosote, selling for fifty cents a gallon, into a motor fuel which would sell for twenty cents a gallon. Coos Bay coal is particularly rich in creosote constituents, consisting mainly of phenols, aerosols, and exylenols and their homologues. Creosote is more toxic and superior to that usually marketed from gas works. At the same time the United States imports more than fifty percent of this substance, and the market is expanding. After these volatile products have been distilled off, there remains the great bulk of original coal, such as char, which has a higher heating value per pound than the coal from which it was made, due principally to the percentage of water driven off. This char is easily ignited. It would be an admirable fuel were it not for a physical structure that is light, porous, dusty, and easily broken in handling. Hence to make a commercial fuel from the char,*

it should be briquetted or prepared in some other manner, which would make it suitable for marketing.[4]

E. W. McInturff and W. T. Stoll along with other men, agreed with Hillstrom and applied for a patent to set up a workable retort for the area.[5]

Dwelling on the possibilities in coal, W. G. "Bill" Robertson, partner of Louis Simpson in the Empire Development Company, commented in the same *Coos Bay Times* article:

Coos Bay coal is not only equal of any coal west of the Rocky Mountains but it is rich in chemicals, gas, tar, and other by products of the highest commercial importance. Scientific experimentation has taught us that in each ton of Coos Bay coal there are 2,800 cubic foot of gas, 500 pound of coal tar products, and about 1,000 pounds of pure carbon, a substance which can, at small cost, be manufactured into a fuel the equal of anthracite, about ninety percent carbon.

Robertson seemed somewhat overly ambitious in his assessment of Coos Bay coal possibilities. Referring to a projected carbonization plant being considered in North Bend, he continued:

Anthracite has sold on the Pacific Coast at prices ranging from $24 to $40 a ton, and charcoal as high as $40 a ton. It is estimated that the fuel manufactured at the North Bend plant will cost less than two dollars a ton, and that seven by-products will be extracted from each ton of coal for a value of $60. This new fuel is the practical answer to the Pacific Coast's fuel problem in the manufacture of steel and iron, an industry at present seriously handicapped in its development by the high costs of anthracite and coke.[6]

In spite of all their plans, the carbonization plant never materialized.

Periodically, dreams and enthusiasm for developing Coos County coals have continued to surface, especially during times of business recession, or national emergencies such as World War II. Because of the war, Coos County officials became concerned with possibilities of a fuel shortage in the area. Judge E. L. Peterson, Judge L. D. Felsheim, along with the county commissioners, Lars P. Peterson and P. W. Culver, petitioned the

Oregon Department of Geology and Mineral Industries to make a study of the coals in Coos County. The state matched the county's $20,000 and in 1943-44 the Oregon Department of Geology and Mineral Industries carried out a comprehensive survey.[7]

The agency started the work in April, 1943, with John Eliot Allen, geologist in charge, and Ralph S. Mason, as mining engineer. In June, Ewart M. Baldwin became assistant geologist, and in October, Francis Cleaver was appointed the engineer. L. L. Hoagland, assayer and chemist for the department, made the coal analyses. The men inspected, surveyed and mapped areas and mines. They took samples for assaying, carbonization, and hydrogenation in their concentration on some 150 out-croppings and mines in the area.[8] The engineers and geologists indicated that both processes could easily be applied to the local coals.

Low temperature carbonization of coal is a heat treatment, in the absence of air, at temperatures of 450 to 700 degrees centigrade. High temperature carbonization is at 900 to 1200 degrees. The process produces a smokeless, easily ignitable, solid fuel. It simultaneously recovers a number of liquid by-products. Briefly, the operators place 250 grams of coal, crushed to pass a 10-mesh screen, in a specially designed aluminum retort. To this retort engineers connect in a series, an ice-cooled tar receiver, a condenser, a glass tube filled with activated charcoal to entrap any tar and light oil mists escaping the condenser, and a gas receiver. They heat the retort to a maximum temperature of 550 degrees centigrade and maintain that temperature for two and one-half hours. The distillation products pass from the retort. The oils, tar, and water are caught in the condensing system or the activated charcoal and the gas collects in the gas receiver. Later they separate the water from the oils and tar and measure the quantity products of the assay, namely, char, tar, and oils, and gas. They analyze the gas, determine the constituents present, and calculate its calorific value.[9]

The hydrogenation process treats coal with hydrogen under high temperature and pressure. This process produces the same products obtained from petroleum. The hydrogenation process is expensive. England and Germany have used the process for many years to produce motor fuels and oils. Hoagland, in charge of the tests for the 1943-44 survey of Coos County coal, sent a forty pound sample from the Thomas Mine to the U. S. Bureau of Mines in Pittsburgh, Pennsylvania, for a small-scale hydrogenation test. They hydrogenated the coal for three hours in a 1-liter rotating bomb (a vessel for collecting compressed gases), at a temperature of 430 degrees Centigrade. Initially the pressure, while yet cold, was 1,500 pounds per square inch of hydrogen. Eighty-eight percent of the coal, on the moisture-free basis, liquefied, and twelve percent remained as an unliquefied residue. Of this residue, about five percent was ash and seven percent unliquefied coal. Of the seven percent of unliquefied coal, five percent probably could have been liquefied had a higher pressure of hydrogen been employed. These tests, in the 1940s, indicated that Coos County coals would be satisfactory for full-scale commercial hydrogenation.[10]

The "Coos King" and "Coos Queen" bicycles of the Marshfield Cyclery in the early 1920s confirmed that land transportation was replacing travel by water on Coos Bay. (Stadden Collection)

During the detailed examination of the Coos coal beds the geologists mapped 980 acres on a scale of 200 feet to the inch with ten-foot contour intervals. The crews drilled a total of 9,000 feet in 322 holes by hand and with power drills. They faced (sampled) and measured coal in 107 places and traced outcrops on eight projects by shallow drilling for a total distance of 18,500 feet.

Engineers R. S. Mason and J. F. Cleaver ran the exploration project. They hired the crews, calculated the payrolls, and laid out the daily work for the crews who sometimes worked in different localities. They saw to the transportation of the crews to and from work, maintained the hand and power drills, and supervised the drilling. They also kept the records and finally calculated the tonnage per acre of the overburden above the coal. Additionally they estimated the extent of coals produced from the various sites. Further, they substantiated, verified and compared their finding with those geologists who had made earlier studies of the area.[11] These studies produced detailed findings for prospective developers of coal-based industry in Coos County.

Allen and Baldwin listed the mines and out-croppings involved in their study:

> *Wilcox, Libby, (Newport) Englewood, Reservoir, Flanagan, Southport, Thomas, Maxwell, Henryville, Delmar, Overland, Martin, Carbondale, Beaver Hill, Klondike, Fahy-Muir, Riverton, Alpine, Timon, Gage, McGee, Rouse, Urquhart, Kite, Smith-Kay, Panter, Lyons, Sell, Eureka, Scorby-Mcginty, Fat Elk Creek, Cedar Point, McClain, Bituminous Coal Company, Yokam Point, Big Creek, South Slough, Oldlands, Vey, Empire, Marsters, Lyons, Albee, Hanson, Black Diamond, Lone Rock, Glasgow (Hardy), Gilbertson, Carlson, Willanch, Ward, Wirth, Smith, Lillian, Gunnell, Messerle, Huntley, West, Smith-Powers, Coaledo, Standley, Noble Creek, Newcastle, Sumner, Ferberish, Hall Creek, Lampa Creek, Donaldson, Woomer, Sevenmile Creek, Seven Devils, Archer, Sengstacken, Bunker Hill, Belfast, Caledonia, Davis Slough, "Thirty-Six," Brown Slough, Stainbeck, Bandon, Lakeside, and Remote.*[12]

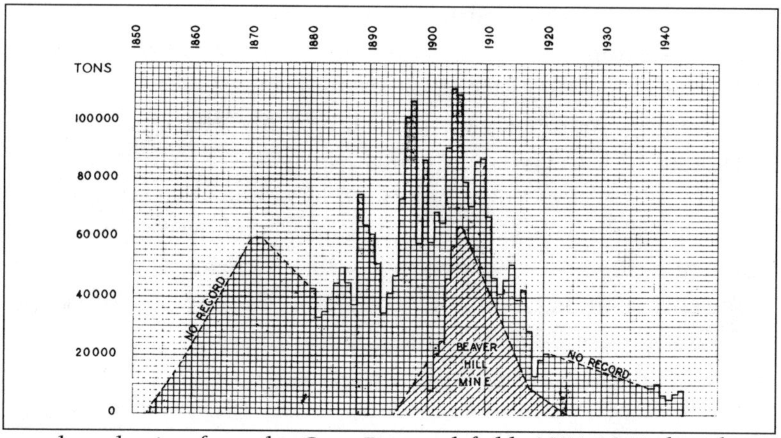

Annual production from the Coos Bay coal field, 1854-1943, based on available records and estimates. (Allen and Baldwin 1944:52)

Both Allen and Baldwin turned their careers to teaching as professors in the Oregon State system of higher education. Baldwin taught at the University of Oregon and Allen at Portland State University. Both taught generations of college students about the geology of Oregon and were living in 1994.

COOS COAL HYSTERIA OF THE 1980S

In the late 1970s it appeared that Japan's energy needs, using Coos County coal, might significantly affect the economy of southwestern Oregon. The Japanese showed signs of changing over to coal for energy. The possibility created high interest because of an unemployment rate of over 15 percent in the county. Many thought that opening up coal production could help the local economy. "It all began with the Japanese" stated Jack Beebe, Coos County Commissioner, referring to the excitement and enthusiasm about possible return to coal production in Coos County, in the early 1980s.[13]

"What happened," Beebe continued, "the oil companies had so increased the price of oil that the Japanese began to feel the pinch for high cost of energy. They told the oil companies that they would begin changing their factories over to the use of coal."[14] In the 1970s the oil producing countries in the Mid-East formed OPEC, an oil cartel. OPEC countries, in their handling of the oil supplies, had actually created the situation. They continued to raise the price of oil to the refineries.

To bring emphasis to their statements, the Japanese began negotiating with coal companies around the Pacific rim, including Australia, the United States, and Canada. Two companies in Vancouver, British Columbia, spent a considerable amount of money building coal-exporting facilities in expectation of shipping coal from that country.[15] The Port of Portland, Oregon, spent $60 million for a facility during that time in anticipation of receiving coal from Wyoming and shipping it to Asian markets. The Port of Portland later found that the expenditure was premature.[16]

The United States, with its millions of automobiles and other energy users such as factories, became heavily dependent on the Middle East oil. Prices rose until they reached a "benchmark" of $36 per barrel. In 1981, some countries charged as much as $42 per barrel. OPEC, of course, was not without its problems.[17]

Some countries, such as Saudi Arabia, the largest producer for the United States, tried to get its associates to hold the line on prices but had some difficulties. At a meeting in Geneva, Switzerland, in August, 1981, producers could not agree on the price. Humberto Calderon, Prime Minister of Venezuela, angrily left the meeting. Venezuela produced large quantities of oil processed in the United States. Saudi oil minister, Sheik Ahmed Zaki Yamani, announced that he would not raise the prices. Already the cheapest in OPEC, he said he would cut back production as much as one million barrels per day. Yamani called the cutback a gesture of good will to other producers having trouble selling their oil for prices up to $42 per barrel. The Saudis, who had been charging $32 per barrel, had been producing 10.3 million barrels a day to glut the market. They were trying to force the others to comply with lower prices.[18]

In 1978 GCO Minerals, a Houston, Texas, firm began negotiations with the Coos County Commissioners for a lease on the coal beds under the Coos County Forest. Canasia, a subsidiary of American Coal Company of Utah, in the early 1980s, made enticing proposals to the Coos County officials to lease the same property. In their talks with officials and others around the county they indicated that the latest methods of mining would be used.

These proposals triggered political ramifications for the county officials. They meant to conserve the value of Coos County's coal beds for the benefit of the citizens. Also, as the commissioners explained, their responsibilities were directed by state statute. They considered the best course was some kind of bidding process. Many people in the area were out of work. Some thought the county should make it easy for a company, any company, to come in and develop a coal facility. The commissioners, however, sought the best results for the county. Three years passed during these negotiations, and in 1981, Canasia said it could not possibly meet the suggested restrictions and conditions of the lease. Canasia stated that the company had worked for three years and campaigned fervently to get the leases through negotiations.[19]

During this time the business recession was seriously affecting the economy of Coos County. Some thought that Jack Beebe, Robert Emmett, and Ed Stevenson, the county commissioners, were too restrictive in their demands. GCO also sought other coal on which the county held the mineral rights. These beds lay under timberlands owned by International Paper Company. Canasia indicated that Consolidated Coal Company would have been a partner in their coal producing activities. Canasia had a lease on the Eden Ridge coals but desired the beds owned by Coos County to meet operation projections. The company planned to bring in coal from Utah, mix with the Coos Bay coal, and ship to Japan.[20]

Attorneys James B. Bedingfield and Jerry Lesan represented other interested companies. They declined to disclose the names of their clients. Some citizens thought that petroleum companies were interested in obtaining leases only to "sit" on them. The commissioners, however, thwarted any such idea by their restrictive lease agreements.[21]

Bedingfield and Lesan, along with the representatives of Canasia, told the commissioners that their bidding requirements were too strict. One of the suggested requirements was the production of a minimum of one million tons of coal in five years.

Commissioner Beebe said, "I'm interested in what's going to benefit the people of Coos County."

Bedingfield replied, "To put this five-year, flat-out production requirement in the lease is something that's just unacceptable to most large companies."[22]

The commissioners then agreed that they would consider a proposed lease from one of the companies as a pattern to follow. They reminded the participants, however, that they were required by the law to make the situation competitive in the interests of the county.[23]

The commissioners worked out the legal problems, established acceptable wording in some leases to meet their fiduciary responsibilities, and began doing business with coal companies. On October 30, 1981, they passed a resolution *(right)* regarding the leasing of the mineral rights to county property. [24]

The signed leases contained many more details than stated in the resolution, defining the necessary duties and restrictions. On November 16, 1981, the county leased 4,258.58 acres to GCO Minerals. April 13, 1982, the county entered into a similar lease with American Coal Company of Utah.[26] The commissioners only leased a portion of the county coal beds. They purposefully held back some of the best coal beds and so informed the leasing companies. The commissioners wanted to see their performance before leasing additional lands to them.[27]

The American Coal Company's plan with Consolidated Coal Company fell through in September of 1981 because Canasia did not obtain the necessary leases.[28] Yet they continued to negotiate and ultimately obtained a lease.

During the period of negotiations Jack Condor, representative for Canasia (American Coal Company), revealed that Canasia was dealing with one of the strongest energy companies in the United States. Their lease on the Eden Ridge coals, south of Powers, provided a potential fifty million tons of recoverable coal. They planned to mine that area at one and a half million tons per year in conjunction with the operations beneath the Coos County Forest. Condor, in a question-and-answer meeting with the public stated:

'THIS MATTER HAVING come before the Board of Commissioners sitting for the transaction of County business on the 30th day of October, 1981, and the Board of Commissioners having previously called for proposals to lease County mineral rights, more specifically coal rights, and GCO Minerals Company having submitted such a proposal on October 9, 1981; and

WHEREAS, the Board of Commissioners and GCO Minerals having entered into negotiations in recorded executive session and the parties having an agreement for the County to let to GCO Minerals the right to mine and sell coal belonging to the County according to the following terms:

1. A one time bonus of fifteen ($15.00) dollars per acre will be paid to Coos County.

2. GCO Minerals will pay rent in the amount of three ($3.00) per acre annually, except that each advance rent payment shall be deducted from any royalty due Coos County upon production occurring during the year for which the rent has been paid.

3. For coal mined and sold to a third party not affiliated with GCO Minerals, Coos County will be paid royalty in the amount of the greater of (1) seventy-five cents ($0.75) per ton of 2,000 pounds or (2) six (6) percent of the average sale price per ton for the coal sold f.o.b. preparation plant or f.o.b. the mine if there is no preparation plant.

4. GCO Minerals shall apply for all permits necessary for conducting coal mining operations within five (5) years following execution of the proposed lease, and shall initiate coal mining operations within ninety (90) days following the issuance of such permits.

WHEREAS, the proposed lease is subject to the Board of Commissioners holding a public hearing to give the citizens of Coos County the right to be heard regarding this proposal;

NOW, THEREFORE, BE IT RESOLVED that Coos County intends to enter into a coal lease with GCO Minerals Company and therefore calls for a public hearing to be held on the 5th day of November, 1981 at 10:00 A.M. to explain the proposed lease to the public and receive comments thereon.

DATED THIS 30 day of October, 1981.

BOARD OF COMMISSIONERS
Robert Emmett
Doc Stevenson[25]

The coal will be trucked from Eden Ridge to the North Spit. There may be as many as 100 trucks a day which will require highway user fees of $10,000 per year. The North Spit facility will contain a coal dust containment system. . . . Coos County coals will be mixed with the higher grade coals from Utah to meet the requirements of the customers.[29]

The leases expired in 1985 after the companies failed to meet the lease terms.[30]

Coal is not a renewable resource. Nothing was indicated in the leases regarding the percentage of recovery from the coal beds. Conventional mining in the past produced only a 60% recovery.[31] Future mining technology could change that figure to nearly 100%. With the public's concern for conservation of natural resources, future contracts or leases could include requirements for a much greater recovery than 60%. Coos County contains over a billion tons of coal but the legacy of mining in the East and its harmful impacts on the environment have led to stringent controls on such activity and reclamation of mined areas.

The completion of the 800 mile oil pipeline across Alaska from the North Slope to Valdez on Prince William Sound took away Canasia's potential customer. Japan took options for millions of gallons of the Alaska oil. Within a few months, the OPEC countries stabilized their crude oil prices.

During the time of the high interest in the coal, plans for developing facilities on the North Spit met with a number of considerations in economic growth and environmental concerns. The Port of Coos Bay considered a number of plans for the area north of the bay. Coos County Planning Director, Bill Grille, announced a $46,000 grant from the Coastal Energy Impact Fund. Grille said that the money could be used for one of three purposes—to hire a coal coordinator who would work with the Planning Department or the Board of Commissioners, to hire a coordinator to work with the Port of Coos Bay, or retain a consultant to deal with the issues of the North Spit. He also mentioned that since Canasia had its own environmental assessment staff a study by a private company might be best.[32]

Bruce Laird, on the marketing committee for the Port, and later president of the Port of Coos Bay, indicated that the Port had made every effort to comply with those interested in developing a shipping facility on the North Spit. He pointed out that the area was fortunate in not having spent a lot of money before the situation was more mature. "We should thank our lucky stars that this mining thing ended before we went that far," he said, referring to the $60 million expenditure by the Port of Portland.[33]

Canasia spent $105,000 in payments to the Port of Coos Bay for a shipping site on the North Spit. It discontinued its payments to the Port; the lease expired on February 10, 1984.[34] Local timberlands in which mineral rights were owned by the timber companies were also under lease for a few years to the energy companies. Soon the economy improved and gasoline prices at the pump stabilized. Within weeks the interest and excitement in the Coos coals subsided. The companies met a few payments and then let their leases go.

The Coos Bay Quadrangle and other areas of Southwestern Oregon hold only a small portion of the thousands of acres of coal in the United States. Nevertheless, the principal drawbacks of the local subbutiminous and lignite coal, high moisture content and friability, may prove to be an advantage in the new technology.[35] Geologists estimate that Coos County has one billion tons of recoverable coal. Although a billion tons seems a lot, officials of the company that Jim Bedingfield represented during the months of negotiations indicated to him that all of Coos County coals would equal only a fraction of the production in the state of Pennsylvania.[36]

THE SUPER FUEL

Dreams sometimes become realities when the entrepreneur takes a few more steps. Rudy Hillstrom was very close in 1930 to obtaining the goal. Hillstrom knew about hydrogenation and carbonization, but it took a chemist to find the magic. One was Douglas V. Keller, Jr. In 1972, Douglas Keller, a professor of material science at Syracuse University in New York, and an engineer, Clay Smith, took hold of the dream and developed a process for a superior liquid fuel from coal. Professor Keller, working with one of his students at Syracuse University, made the discovery. Keller and Smith were fellow club members and they were discussing the experimentation of the student. Much work remained to refine the process, but Smith got excited when he heard the results.[37]

The student, Robert Aldrich, had been searching for a way, or some agent that would make it easier for drills to penetrate coal. Using Keller's suggestion, the student changed from acids to an alcohol base. He applied ammonia, Keller explained, and the raw coal broke down into fine particles. The agent separated the purer hydrocarbons from rock and pyretic sulfur and significantly reduced the ash content. Aldrich later became vice-president of the Electric Power Research Institute, a consortium of companies developing new energy ideas. Clay Smith was an engineer who had installed skating rinks and other refrigeration projects.[38]

Two years before, in 1970, Congress passed the Clean Air Act. Environmental concerns were highlighted and publicized during the nation's first Earth Day. Environmental awareness increased and clean-burning fuels became the concern of the public.[39] Sulfur Oxides (SO_x) and nitrogen oxides (NO_x) produced in the burning of coal are the principal causes of acid rain. More than half the nation's electricity is produced by coal power plants. They burn coal at the rate of several billion tons per year.[40] The time seemed right for Keller and his clean-burning fuel for power companies.

Much work remained in refining the product to meet their aspirations and provide the ideal fuel. Keller and Smith formed a company and applied for a patent on the process. They adopted the Iroquois word "Otisca," meaning "water that has gone away," as a name for their process and their company.

The two men approached the project with carefully planned scientific methods. They rented a place for a laboratory and went to work. They carefully documented all the details of their process, even to the minute size of the powdered coal, and the measured results. They hired four chemists to help them and spent the better part of fifteen years developing the product. They milled and tested over five hundred different samples of eastern and midwestern coals. They noted all the characteristics of the various samples of coal and made a detailed record of the results.[41] Keller and Smith became knowledgeable and soon gained the reputation for producing the best of liquid fuels from coal.

Not only was it a cleaner burning fuel, Otisca's liquid fuel was easier to transport than coal. Its advantages in terms of transportation and use in engines, including diesels and turbines, appeared to put it in direct competition with oil and gas.

All that work was not simple. They primarily ran finely ground coal through a chemical bath. The process separated out all but 1% of the mineral content, or ash, and 0.5% of the sulfur that formed sulfur oxides. Because it is half water, Otisca fuel produces a cooler flame than straight coal and hence about half the nitrogen oxides. They used ammonia first, then switched to the refrigerant, Freon, as the separating agent. Later, they found Pentane more satisfactory. Pentane is one of the three isometric hydrocarbons of the methane series occurring in petroleum.[42]

In January, 1988, Professor Keller wrote an introduction to a comprehensive report of their activities in demineralization of coal.

> *The demineralization of coal by selective agglomeration using pentane as an agglomerant which has been called the "T-Process" involves two steps. First, the raw coal is reduced to some particle size smaller than 0.5 mm. And secondly, the fine coal is recovered from a dilute raw coal water slurry by the dispersion of pentane in the slurry. The pentane envelops the coal particles which allow the formation of liquid bridges between the coal particles that results in a water immiscible mass of coal and pentane. The latter is separated from the dispersed mineral matter in the water phase by some simple mechanical process. The coal recovery from very fine raw coals is normally in excess of 95%. The pentane is recovered by evaporation condensation for a reuse in the process leaving the demineralized coal for utilization either as a slurry or in dry pellets. Pilot opera-*

tions using this process have produced over 125 tons of pyrite-free product coal with an average ash content in the range of 0.8 wt%. . . .[43]

Keller further explained the process:

The demineralization of coal utilizing the selective agglomeration with pentane, i.e., the Otisca T-Process, involves two simple steps which consist of particle size reduction of the raw coal to below 600 micrometers and separation of the coal-bearing particles from the mineral matter particles that were released by fracture in a dilute water slurry by the addition of liquid pentane to effect the separation. Pentane (C5 H12) is immiscible in water and with proper dispersion in the water/coal/mineral matter mixture the pentane coats the coal particles providing liquid bridges between the coal particles such that they bind together forming a mass which is readily separated from the mineral particles that remain dispersed in the water. The liquid/liquid/solid separation process has a long history in its own right and particularly so in reference to its application to the demineralization and recover of fine coal.

The purpose of the separation process in the demineralization of coal and the extent of demineralization is dependent on the application of the product coal . . ..

The report provided application data in the processing of over 100 different coal seams.[44]

The 1970s were promising years for the new energy development plans. Soaring oil prices prompted industry to search seriously for alternative energy sources. Otisca did the first pilot project with Island Creek Coal Company, a 15-ton-per-hour operation in Bayard, West Virginia, at the headwaters of the Potomac. Otisca also did some early business with General Public Utilities in western Pennsylvania. Later, they worked with other companies such as Florida Light and Power, Westinghouse, and General Motors. General Motors developed an experimental coal-powered Cadillac that ran on Otisca fuel. They called the car *Coal-dorado*. Five big companies—General Electric, Norfolk Southern Railway, Eastern Fuels, Westmoreland Coal and Zurn Industries, jointly invested $8 million with Keller and Smith in November of 1984.[45]

It seemed at last, that Otisca was off and running, but such was not to be. Complications continued to haunt its activities. Several dozen employees, burgeoning paperwork, outside investors with varying interests, and government requirements caused devastating frustrations.[46]

In the spring of 1986, world oil prices dropped below $15 per barrel. The impetus to seek out alternative fuels waned. Florida Power & Light left, joined a regional grid in the South, completed a nuclear plant, and signed a contract to buy cheap oil from Venezuela.[47]

In 1988 Otisca won a Clean Coal matching grant from the Department of Energy. A $7 million grant was to pay for changing several Syracuse-area boilers to use Otisca fuel. The grant required a third of the money from the private sector. Smith and Keller immediately asked their preferred stockholders for cooperation. To their consternation, however, the matching money was not offered. General Electric, the largest and most influential of the five, was more interested in the locomotive business instead of fixed boilers. The Energy Department office in Pittsburgh that sponsored Otisca's application promoted experiments for plants using fixed boilers.[48]

For one reason or another the companies refused to cooperate on the matching funds, except Zurn Industries. Zurn made boilers for industrial use. Zurn, however, was not prepared to back the grant alone. Without matching funds, to their dismay, Keller and Smith lost the grant. Soon, however, another company, CSX Corporation, parent of the Chesapeake & Ohio Railroad, took an interest in Otisca. CSX planned a cogeneration plant at the Greenbrier Resort in West Virginia. As the nation's largest mover of coal, CSX had an interest in promoting its use and export. CSX found that Otisca was the only firm capable of producing a coal-based liquid fuel that met their specifications.[49]

The Greenbrier project would have been ideal to demonstrate the new fuel, but the Department of Energy had decided to work with other power plants and technologies. Thus, the entrepreneurs at Otisca experienced frustration almost beyond reason, especially since they were ahead of others in that energy field.[50]

In 1994 Keller indicated that Otisca fuel costs approximately 60 cents per million BTU.[51] Adding mining and other production costs for a conservative maximum of $3.00 per million BTUs suggested the product was a bargain compared to $4.50 per million BTUs for natural gas, or $10 per million BTUs for gasoline. In explaining the reluctance of companies to make changes, Keller said that the average energy costs ran approximately 10% of their budgets. CEOs and corporate board members are aware that savings in production costs become profits. However, their concerns for larger money matters and profits motivations obscure the few percentage points that might be saved by changing the energy source. Also, such savings might not be evident until costs of changing are amortized.[52]

Even with all its superior fuel qualities, power companies and others consider that Keller's fuel is really not proven until actually placed on the line in one of their plants. Their reluctance to start using the new fuel, partly caused by the cost of changing and of other factors big and small—lend to a "catch 22" situation. Until such action is taken, the acceptable "proof" will not be forthcoming. Electric power companies operate under laws which force any gain above certain economic levels to be passed along to the consumer of their products. Thus, if they spend money developing a system that fails, they are not compensated. If the development is successful they have to share with their customers. Keller, in thinking of the frustrations, has said, "If you're not an optimist, you're out of the game," confident that someday, and perhaps at almost anytime, the golden dream will be complete.[53]

The brief resume of the Otisca story could point the way to Coos County's future. Technology of today suggests the potentials for the development of mining far beyond the wildest dreams of the pioneer coal miners of southwestern Oregon. Although Coos County has been quite thoroughly studied and surveyed for its coal beds, new techniques could be important in quickly determining the degree of slant of a vein of coal, or the thickness of the bed, or other features affecting future mining.

A recent development allows the engineers to "read" the various layers in the earth. Ground Penetrating Radar (GPR) sends high-frequency electromagnetic impulses into the ground. When the signal hits the interface between two subsurface layers, part of the signal is reflected back. These signals are transformed into a visual display that provides an immediate view of the data. A stylus on a graphic recorder sweeps across a paper at a uniform speed. The reflected signals cause the paper to be darkened at points in proportion to the strength of the reflection.[54] GPR holds the potential for an even more detailed reading of Oregon's coal deposits.

Modern mining methods of conveyor chains and other advanced mechanical techniques of digging and bringing coal to the surface may some day be replaced with even more advanced engineering. For example, Canadians at Smallwood, British Columbia, pump water into the coal beds. They make the coal into a slurry that they pump back to the surface. The mines at Black Mesa in Arizona use the same technology.[55] These engineering techniques do away with human presence underground and could lower production costs at the adit. They introduce a host of new environmental concerns, especially related to water quality and possible contamination. Coal is now in a position to challenge higher petroleum prices.

Other engineering techniques and methods are being encouraged by the Department of Energy. One of these is a process called Magnetohydrodynamics, or MHD. The process is founded on Michael Faraday's discovery in 1831. Faraday found that moving an electric conductor in a magnetic field created an electric current. The MHD process forces hot gases through superconducting magnets to do the same thing. Traditional coal-powered electric plants convert only 32 to 35 percent of the heat into electricity from burning coal. With the addition of the HMD process to their plants they are able to obtain as much as 60 percent. In 1993 owners of a plant in Billings, Montana, drew plans to build such a facility.[56]

There are still other coal energy development plans; one is cell power. Quite simply it is as follows:

The raw coal enters a conventional gasifying plant, which produces a fuel gas at between 650 degrees Centigrade and 1,000 degrees Centigrade. This gas passes through a cleanup process, removing extraneous particles, sulfur, nitrous oxide and other substances. The cleaned-up gas then passes to the anodes of a stack of fuel cells, which also consume air, to produce direct current electricity. Hot exhaust gas, mainly carbon dioxide and water, from the fuel cells' cathodes creates steam to drive a turbine, producing more electricity. The overall efficiency could be more than 45 percent.[57]

Manufacturing products from the chemicals could prove to be more important for Oregon coals than extracting energy. Scientists and industrialists are looking to coal for chemicals to make plastics, solvents, fertilizers and many other products. Coal contains the hydrocarbons that are the basic molecules for the world's chemical industry. Chemists improving on the Fischer-Tropsch Process use a number of catalysts to "select" the hydrocarbon molecules. One such catalyst is nickel.[58] Facilities for importing nickel through the port of Coos Bay could hold the potential for use of the Fischer-Tropsch, or a similar process, on southern Oregon coals.

Black crude oil may, in the near future, be confronted with a competitor that more easily meets environmental requirements or provides product chemicals. The world's oil and gas are limited to approximately another 50 years at the present use.[59] Therefore, with coal as the most abundant fossil fuel the results are almost a predictable leadership in power production. The several billion tons of coal lying in southwestern Oregon may again be a substantial part of the region's economy.

ENDNOTES

1. Emil R. Peterson and Alfred Powers, *A Century of Coos and Curry*, pp. 458-459.
2. *Coos Bay Harbor*, North Bend, Oregon, June 7, 22, and October 4, 1907.
3. Peter Maitils and Jon Rourke, Rich Chemicals, *New Scientist*, January, 1993, p. 37.
4. *Coos Bay Times*, Southwestern Oregon Edition, January 30, 1930, p. 15.
5. Ibid., pp. 15-16.
6. Ibid., p. 15.
7. John Eliot Allen and Ewart M. Baldwin, *Geology and Coal Resources of the Coos Bay Quadrangle, Oregon 1943-1944*, p. 3.
8. Ibid.
9. Ibid., p. 60.
10. Ibid., pp. 59-60.
11. Ibid., p. 53.
12. Ibid., pp. 64-65.
13. Jack Beebe, and Ed "Doc" Stevenson, Interview with Dow Beckham, September 14, 1992
14. Ibid.
15. James Bedingfield, Interview with Dow Beckham, September 14, 1992.
16. *The World*, Coos Bay, Oregon, February 16, 1984.
17. Ibid.
18. Ibid., February 16, 1984.
19. Ibid., August 18, 1981.
20. Ibid., September 2, 1981.
21. Ibid., September 5, 1981
22. *The World*, op. cit., September 5, 1981
23. Ibid.
24. Jack Beebe and Ed "Doc" Stevenson, op. cit., September 14, 1992.
25. Ibid, Document from the Coos County commissioners, September 14, 1992.
26. Ibid.
27. Ibid.
28. *The World*, op. cit., September 6, 1981.
29. Ibid., August 18, 1981.
30. Ed "Doc" Stevenson, op. cit., September 14, 1992.
31. Allen and Baldwin, op. cit., p.72.
32. *The World*, August 26, 1981.
33. Ibid., February 16, 1984.
34. Ibid.
35. *New Scientist*, op. cit., January, 1993.

36. James Bedingfield., op. cit.
37. Douglas V. Keller, Jr., Interview with Dow Beckham, February 5, 1993.
38. Ibid.
39. Ibid.
40. Ibid.
41. Ibid.
42. Douglas V. Keller, Jr. and W. M. Burry, The Demineralization of Coal Using Selective Agglomeration by the T-Process, *Coal Preparation,* 1990, Vol. I, p. 10.
43. Ibid., pp. 1-17.
44. Ibid.
45. Douglas V. Keller, Correspondence, January 26, 1993.
46. Ibid.
47. Ibid.
48. Ibid.
49. Ibid.
50. Ibid.
51. Keller interview with Dow Beckham February 5, 1993.
52. Ibid.
53. Keller, Interview with Dow Beckham, January 25, 1993, followed by a letter dated January 26, 1993, giving permission to cite his information.
54. Moshe Kochavi,Timothy Renner, Ira Spar and Esther Yadin, *Biblical Archaeology Review*, The Land of Gosher, July-August, 1992, p. 51.
55. Keller, op. cit., January 25, 1993.
56. Dan Clark, *New Scientist*, In Search of a Better Burn, January, 1993, p. 51.
57. Ibid., p. 24.
58. Peter Maitils and Jon Rourke, *New Scientist,* Rich Seams for Chemicals, January 23, 1993, p. 38. The authors observed: "The most important thing in the chemical process is selectivity—to obtain useful amounts of products which are pure. The key to better selectivity lies in the addition of various "promoters," chemicals which enhance the catalysis. . . ." Some other catalysts are iron, cobalt, and tuthenium.
59. Sue Bowler, *New Scientist*, Where the Power Lies, January, 1993, p. 12.

CONCLUSION

The prospects of coal worked like a magnet to draw investors, speculators, and settlers in the 1850s to the southwest coast of Oregon. The harbor at Coos Bay, in spite of its shifting channel and perils for sailing ships, beckoned to brave mariners who risked their ships and lives of their crews to carry cargoes of coal to the markets of California. The "Age of Steam" with locomotives and boilers to power sternwheelers and sidewheelers created a demand for West Coast coal. The seasonal fluctuations in temperature in the communities around San Francisco Bay likewise fostered the market.

Often inexperienced and driven more by high hopes than a realistic assessment of the coal deposits, the investsors in the mines scrambled for a hold on key properties and transportation linkages. A number—such as Utter and Ojeda—simply selected the wrong site and wasted their energies and backers' capital. Some—such as Patrick Flanagan and Samuel Stilman Mann—had "lady luck" on their side and emerged with prosperous operations. The Newport Mines at Libby were blessed by consistent deposits, good drainage, and relatively easy transportation of coal via a trestle down Coal Bank Slough to bunkers on Coos Bay.

Throughout the nineteenth century, however, the mines of Coos Bay and Coquille region were beset by problems. The markets for coal remained uncertain and seasonal. Competition from the Mount Diablo coal fields in California, from the Dunsmuir mines on Vancouver Island, and from the mines at Black Diamond, Washington, sometimes proved ruinous. The lack of major harbor improvements—jetties and dredging—rendered the Coos Bay bar a foe which might, with the slightest dip in the wind, turn a profitable voyage into a disaster.

Laborers found that the Coos County coal mines afforded a begrudging living. Most faced weeks, if not months, of unemployment. Thus the miners and their families embraced a subsistence lifeway. They tended vegetable gardens, kept livestock and poultry, hunted deer, elk, ducks, and geese, fished and dug clams, and worked diligently to preserve the foodstuffs to sustain their children throughout the year. They had no minimum wage, protection of safety, or even reasonable limits on hours or holidays. Boys as young as thirteen or fourteen left school to enter the mines just as others worked above ground in logging operations as "whistle punks" and, when older, as choker setters. The fare was grinding, unrelenting work. When the mines closed with a downturn in sales or profits, the scramble to survive became even more frantic.

The coal fields of Coos County held their own for nearly sixty years but ultimately succumbed to the advent of internal combustion engines and the shift from steam to diesel and gasoline. Steadily the demand for coal diminished. Mine owners closed their company stores and in time, sold off the clusters of shacks on the hillsides near the portals. While many retained their mineral rights, they let the surface lands go to the county for non-payment of taxes in a begrudging acknowledgment that times had changed.

The alders and firs grew back along the ridges and down the slopes. The slippery clay banks gave way, cutting off old portals and, on occasion, the mountain simply closed in on the cavernous gangways and "rooms" which stretched beneath its surface. Dark settled permanently in the mines. No lantern or head-lamps broke the chill blackness. Only the drip of water, the creaking of the massive blocks of sandstone resting on the shoring and shims, and the rising murky flood which drowned the routes where boys, men, and mules had labored for years marked the course. No longer did workmen struggle with picks, carts, and frustrated dreams. And no one implored God: "Fling us a handful of stars!"

The stars winked in the heavens above to light the nights of new generations in coastal southwestern Oregon whose labors focused on logging, lumbering, transportation and trade. Knowledge of the coal deposits yet stimulated the imagination, drawing in the geologists, Allen and Baldwin, at the advent of World War II to assess the prospects for renewed mining. Energy shortages in the 1970s again sparked widespread interest but died with the declining oil prices and an end to the Mid-East embargo which, for a time, challenged American industry and mobility.

In the 1990s the massive coal deposits remain—a time certificate of sorts. They are nature's bank account of captured energy resource insecurely held and almost within grasp. Chemical treatments such as that proposed by Douglas Keller may yet lead to further development of the mines. It may not. The subbituminous coals of southwestern Oregon face intense competition as an energy resource. Their future is unknown. The history of their past development, however, reverberates like an echo resounding in the deep shafts plunging beneath sea level in Beaver Hill Mines. Memories and voices—a bit whispered—speak of what once was. They hint of what may someday will be.

APPENDIX A

GLOSSARY

Adit. Entry to a mine that slopes slightly upward thus allowing for natural drainage of water. Many refer to any mine opening as the adit.

Aerosol. A suspension of fine solids or liquid particles in gas.

Afterdamp. Mine air with a high concentration of carbon monoxide following a mine fire or methane explosion.

Agglomeration. The action or process of collecting in a mass.

Anticline. An arch of stratified rock or coal in which the layers bend downward in opposite directions from the crest; opposed to Syncline.

Anthracite coal. A hard natural coal of high luster differing from bituminous coal in containing little volatile matter.

Arles. A Scottish term used to designate bounty coal or extra pay.

Assay. Ore analysis or examination to determine the characteristics of a substance.

Barkentine. A three-masted ship having the foremast squared-rigged and the main mast and the mizzenmast fore-and-aft rigged; often referred to as a bark. A bark, according to Webster, is a three-masted ship with the foremast and main-mast square-rigged and the mizzen-mast fore-and-aft-rigged.

Bituminous coal. A coal that yields considerable bituminous matter when heated.

Bitumen. Any of various mixtures of hydrocarbons often together with their nonmetallic derivatives.

Brattice. Curtains made of sacking or canvas which direct the flow of air from the mine ventilation system into places where gas can collect such as at the face or in potholes in the roof.

Brig. A two-masted, square-rigged-ship.

Brigantine. A two-masted, square-rigged ship differing from a brig in not carrying a square main-sail.

Bunker. A wood structure at the docks where the coal was stored before loading into ships.

Butane. Either of two, isomeric, flammable, gaseous, paraffin hydrocarbons.

B.T.U. British Thermal Unit; the quantity of heat required to raise the temperature of one pound of water one degree Fahrenheit at or near 39.2 degrees F.

Carbonization. A process of heat treatment of coal to determine its various elements.

Carbide. A compound of carbon and calcium.

Carbide lamp. Water dripping on the carbide produces acetylene which is lit with a flint inside a glass cylinder.

Carpal tunnel syndrome. An injury common to people who put their hands through repetitive motions such as typing. The repetition and pressure on the wrists can cause swelling and irritation in the tendons that connect the muscles and bones in the wrist. This area is known as the carpal tunnel The swollen and inflamed tendons squeeze the midian nerve which results in a burning pain and numbness in the thumb, index and middle finger.

Catalyst. A substance or an agent that initiates a chemical reaction and enables it to proceed under milder conditions than otherwise possible.

Char. Charcoal.

Check weighman. A miner who checks that the men are credited with the proper weight when the cars are weighed and unloaded.

Chokedamp. Also called blackdamp; mixture of carbon-dioxide with either air or nitrogen; heavier than air; can kill by explosion or asphyxiation.

Clinker. A hard mass of fused, stony matter formed in a furnace or boiler from impurities in the coal.

Cog. Four-sided structure built of timbers and filled with waste rock used to hold up a dangerous roof.

Collier. Coal miner.

Creosote. Brownish liquid consisting chiefly of aromatic hydrocarbons obtained from distillation of coal tar. Also a phenolic compound obtained from wood tar.

Crosscut. Tunnel joining two parallel levels; a pillar of coal is left when two levels are joined by two crosscuts.

Darg. Concept of a "day's work" used in Scotland and Northern England based on how much the average man could dig per day in a given location.

Damp. Any of a number of gases found in coal mines.

Digger. The coal miner at the face digging coal.

Door boy. Boy employed to sit by a light stopping, open it when necessary, and make sure it closed again; also see trapper.

Drawing gas. Indication that a safety lamp is not airtight and it is possible for explosive gas to reach the flame inside.

Drift. A horizontal passageway driven into or along the path of a coal seam.

Dross. Coal dust regarded as harmless until its explosive characteristics were discovered, and useless until it was employed for making coke.

Face. Leading surface of the mine workings; the wall of coal which the miner worked to advance his stall or the tunnel.

Fault. A fracture in the earth's crust accompanied by a displacement of one side of the fracture with respect to the other and in a direction parallel to the fracture.

Firedamp. A mixture of mine air and methane released when the coal is loosened; lighter than air; cause of many mine explosions.

Friable. Coal which is easily crumbled or pulverized.

Freon. A trademark used for any of various non-flammable gaseous and liquid fluorinate hydrocarbons as a refrigerant and as a propellant for aerosols.

Fore-and-aft-rigged. A sailing ship in which most or all of the sails are not attached to the yards but are bent to gaffs or set on the masts or other stays amidships in line with the vessel.

Gob. Area of a mine where coal has been removed and where waste rock is dumped.

Hydrogenation. To combine or treat with or expose to hydrogen.

Immiscible. Incapable of mixing or attaining homogeneity.

Lagging. Rough lumber or slabs laid on stringers and fastened to timber to line roof and sides of tunnels.

Level. Horizontal passageway in a mine.

Lighter. Large open barge which can be brought alongside a ship.

Methane. A colorless, odorless, flammable, gaseous hydrocarbon that is a product of decomposition of organic matter in marshes and mines.

Nickel. A nearly silver-white, hard, malleable, ductile metallic element of a high polish and resistant to corrosion; used chiefly in alloys and as a catalyst.

Out-crop. The place at the earth's surface where a seam of coal breaks through.

Oversman. A foreman.

Pentane. Any of three isomeric hydrocarbons of the methane series occurring in petroleum.

Partings. Layers of clay or other material in the coal vein.

Phenol. A caustic, poisonous, crystalline, acidic compound (C_2 H_2 OH) present in coal tar and wood-tar; often when in dilute solution, used as a disinfectant.

Pillar and stall (room). A method of mining in thicker seams; levels and crosscuts form a grid which helps circulate the air of the ventilation system; the grid leaves behind pillars of coal which support the roof.

Picking table. Sorting bins of screens.

Pithead. Entrance of the mine.

Pothole. Irregularity in the roof where methane gas can be trapped.

Rank. Value of coal or its classification.

Refuse coal. Coal too fine to be usable.

Retort. A vessel in which substances are distilled or decomposed by heat.

Rib. Side of the mine tunnel.

Rope rider. A worker who accompanied the cars of coal out of the mine.

Shaft. Perpendicular entrance to a mine.

Short weight. Occurs when the weight of a car is estimated instead of actually weighed and the miner is cheated by too low an estimate.

Slacking. A tendency for coal to break into small pieces; especially coal with high moisture content exposed for evaporation of the moisture.

Slope. An inclined entrance to a mine; many ordinary tunnels are also referred to as slopes.

Slurry. A watery mixture of insoluble matter.

Square-rigged. A ship with yards and sails running across the vessel as opposed to fore-and-aft rigged where the sails are "in-line" with the vessel.

Stall. "Room" off a slope or tunnel where one or two miners dig coal.

Stoppings. Barriers used to block off openings in tunnels; "heavy" stoppings are permanent barriers constructed of two-inch timber, rock or brick "light" stoppings are wooden doors or curtains that can be opened briefly by trappers or door boys to allow for passage of coal or personnel.

Stringers. Timber that forms the top of the three-sided arch used to support mine tunnels.

Syncline. A trough of stratified rock (or coal) in which the beds tip toward each other from either side, as opposed to anticline.

Tamper. A stick or metal rod used to pack clay, sand or dust around the charge in a drill hole; must not be made of iron or steel because those metals can cause a spark.

Tipple. Structure at the mine entrance for receiving the coal for weighing, sorting, and cleaning.

"To the dip." Following the angle of the seam.

Trapper. Usually a boy employed to sit by a light stopping, opening it when necessary and making sure it is closed again.

Tunnel. Horizontal entrance to a mine.

Upcast. Ventilating shaft of a mine where the air is drawn up to the surface.

Whin. Hard rock, inferior kind of jade associated with drilling boreholes; also called greenstone, whinstone, or whinsay.

BIBLIOGRAPHY

INTERVIEWS

Alpine, Karl
1992-93 Interviews with Dow Beckham, April, 1992; September 1, 1992; March 13, 1993.

Anderson, Bill
1992 Interview with Dow Beckham, July 20.

Barton, Dorothy
1992 Interview with Dow Beckham, July 20.

Baer, Sam
1992 Interview with Dow Beckham, May 15.

Bedingfield, James
1992 Interview with Dow Beckham, September 14.

Briggs, Leonard
1992 Interviews with Dow Beckham, April 20 and September 15.

Beebe, Jack
1992 Interview with Dow Beckham, September 14.

Callery, Martin L.
1993 Interview with Dow Beckham, March 16.

Chard, George
1992 Interview with Dow Beckham, July 20.

Emmett, Robert
1992 Interview with Dow Beckham, July 5.

Flanagan, Dr. Jack
1992 Interview with Dow Beckham, May 10.

Fox, Carol (Archer)
1992 Interview with Dow Beckham, September 12.

Gibbs, Leonard
1992 Interview with Dow Beckham, September 15.

Gilfillan, Homer
1992 Interview with Dow Beckham, March 1.

Gilfillan, John and Edna
1992 Interview with Dow Beckham, October 26.

Gorbutt, Helen (Bomar)
1992 Interview with Dow Beckham, October 15.

Jacobson, "Sticker" Thurston
1992 Interview with Dow Beckham, April 12.

Keller, Dr. Douglas V.
1993 Interviews with Dow Beckham, January 25 and February 5.

Kingsley, Thomas
1992 Interviews with Dow Beckham, April 20 and September 1.

Lapp, Florence
1992 Interview with Dow Beckham, September 9.

Lynch, Dan
1992 Interview with Dow Beckham, October 1.

Macy, Dr. Robert
1992 Interview with Dow Beckham, August 26.

McKelvie, Alex
1992 Interview with Dow Beckham, June 10.

Menegat, John
1992 Interview with Dow Beckham, August 31.

Moore, A. J. "Pi"
1992 Interview with Dow Beckham, October 7.

Phillips, Esther (McGee)
1993 Interview with Dow Beckham, March 3.

Rice, Eleanor (Menegat)
1992 Interview with Dow Beckham August 31.

Rumbaugh, Allan E.
1993 Speech before the Rivermen's Club, Coos Bay, OR., January 25.

Stevenson, Ed "Doc.
1992 Interview with Dow Beckham, September 14.

Stovall, George
1992 Interview with Dow Beckham, July 14.

MANUSCRIPTS

Beckham, Stephen Dow and Rick Minor
1980 Cultural Resource Overview of the Coos Bay BLM District. MS Typescript, Coos Bay BLM District, North Bend, OR.

Bureau of the Census
1900 Twelfth Census of the United States, Coos County, OR., Coaledo Precinct, Beaver Hill, OR., Microcopy T-623, Roll 1346, RG 29: Records of the Bureau of the Census, Washington, DC.
1910 Thirteenth Census of the United States, Coos County, OR., Coaledo Precinct, Beaver Hill, OR., Microcopy T-624, Roll 1280 RG 29: Records of the Bureau of the Census, Washington, DC.

Case, George Baxter
1952 The History of the Port of Coos Bay, 1852-1952. M.A. thesis, University of Texas, Austin, TX.

Coos County Commissioners
1981 Coal Lease, GCO Minerals Company, November 16.

1982 Coal Lease, American Coal Company, April 13. RS 81-022, in the Matter of Leasing Mineral Rights. Coos County Clerk's Office, Coquille, OR.

Davis, Andrew J.
n.d. Accounts and Correspondence, Coos Bay, Oregon, MS, New York Historical Society, New York, NY.

Flanagan, John W
1938 Biography of Patrick Flanagan. Clippings, Coos Bay Harbor (North Bend, OR) Typescript, MS CB-F613, Knight Library, University of Oregon, Eugene, OR.

Fowler, J. E. Z.
1921 Beaver Hill (Preuss) School Census, November 7, Coos County Educational Services District, Coos Bay, OR.

Fox, Carol (Archer)
1922 Biographies of Forefathers: Monroes, Archers, and Brynns, Typescript in possession of Carol (Archer) Fox, Coos Bay, OR.

Gilfillan, Homer
1979 Biography and Personal Recollections of Life in the Early Twentieth Century, Lecture, Coos County Historical Society, North Bend, OR., Typescript, Coos County Historical Museum, North Bend, OR.

Sunset Cemetery
n.d. Burial Records, MS, Sunset Cemetery, Coos Bay, Oregon.

Welch, George
1968 Coos County Coal Mines, Typescript in possession of Dow Beckham, Coos Bay, OR.

West, Victor
n.d. Hardy Mine, and, Coos Bay Union Coal Company, Manuscripts, in possession of Victor West, North Bend, OR.

GOVERNMENT DOCUMENTS

Allen, John Eliot and Ewart M. Baldwin

1944 Geology and Coal Resources of the Coos Bay Quadrangle, Oregon, *Bulletin 27, Oregon Department of Geology and Mineral Industries*, Portland, OR.

Diller, Joseph Silas

1914 Mineral Resources of Southwestern Oregon. *U. S. Geological Survey Bulletin* 546, pp. 1-147.

Libbey, F. W.

1938 Progress Report on Coos Bay Coal Field, *Bulletin 20, Oregon Department of Geology and Mineral Industries*, Portland, OR.

U.S. Life-Saving Service

1897 *Annual Report of the U.S. Life-Saving Service.* Government Printing Office, Washington , DC.

1910 *Annual Report of the U.S. Life-Saving Service.* Government Printing Office, Washington, DC.

NEWSPAPERS

Coos Bay Harbor (North Bend, OR)
Coos Bay News (Marshfield, OR)
Coos Bay Times (Marshfield and Coos Bay, OR)
Coos Bay Times, Southwestern Oregon Edition (Marshfield, OR)
Coquille Sentinel (Coquille, OR)
The Islander (Victoria, BC.)
Marshfield Sun (Marshfield, OR)
Myrtle Point Herald (Myrtle Point, OR)
The Oregonian (Portland, OR)
San Francisco Chronicle (San Francisco, CA)
Supplement to Coos Bay Times (Marshfield, OR), January 30, 1930.
The World (Coos Bay, OR)

ARTICLES

Bowler, Sue

1993 Where the Power Lies, *New Scientist*, 137:12, January.

Charles, Dan

1993 In Search of a Better Burn, *New Scientist*, 137:20-25, January.

DeNeffe, Frederick M.

1956 The Mysterious Shoe-String Railroad, *Oregon Historical Quarterly*, 57:205-228.

Keller, Douglas V.

1981 The Demineralization of Coal Using Selective Agglomeration by the T-Process, *Coal Preparation*. 1990 Vol. 8., pp. 1-17.

Kochavi, Moshe, Timothy Renner, Ira Spar and Esther Yadin

1992 Rediscovered: The Land of Geshur, *Biblical Archeology Review*, Vol. 18, No. 4., pp. 30-44, July - August.

Maitlis, Peter and Jon Rourke

1993 Rich Seams for chemicals, *New Scientist*, 137:37-41, January.

Stone, Boyd

1992 Beaver Hill Mines, *Coquille Sentinel* (Coquille, OR), October 14.

Sullivan, Raymond and John Waters

1980 History of Mount Diablo Coalfield, *California Geology*, March.

BOOKS

Audain, James

1955 *From Coal Mine to Castle*, Pageant Press, New York, NY.

Beckham, Anna M. and Dr. Everett Curry

1979 *The First Baptist Church of Coos Bay (Marshfield), 1879-1979*, Wegferd, North Bend, OR.

Beckham, Dow

1991 *Swift Flows the River*, Arago Books, Coos Bay, OR.

Beckham, Stephen Dow

1971 *The Simpsons of Shore Acres*, Arago Books, Coos Bay, OR.

1973 *Coos Bay: The Pioneer Period, 1851-1890*, Arago Books, Coos Bay, OR.

1987 *Land of the Umpqua: A History of Douglas County, Oregon*,Douglas County Commissioners, Roseburg, OR.

Bowen, Lynne

1982 *Boss Whistle*, Oolichan Books, Lantzville, BC.

1987 *Three Dollar Dreams*, Oolichan Books, Lantzville, BC.

Department of Travel Industry

n.d. *Craigdarroch*, Society for the Maintenance and Preservation of Craigdarroch Castle, Victoria, BC.

Dodge, Orvil O.

1898 *Pioneer History of Coos and Curry Counties, Oregon*, Capital Printing Co., Salem, OR.

Gillespie, Glenn, ed

1985 *The Pacific Power Story*, Lane-Standish Printing Co., Portland, OR.

French, Giles

1964 *Cattle County of Peter French*, Binfords & Mort, Portland, OR.

Gosnell, R. Edward

1939 *The Story of Hatley Park: Residence of the Late Hon. James Dunsmuir and Mrs. Dunsmuir of Pioneer Vancouver Island Family,* Diagen-Hobben, Victoria, BC.

Loftus, Mitzi Asai

1990 *Made in Japan and Settled in Oregon*, Pigeon Point Press, Coos Bay, OR.

Mahaffy, Charlotte L.

1965 *Coos River Echoes, A Story of the Coos River Valley*, Interstate Press, Portland, OR.

Osborne, Ernest L.

1978 *Wooden Ships and Master Craftsmen*, Bandon Historical Society Press, Bandon, OR.

Osborne, Ernest L. and Victor West

1981 *Men of Action, A History of the U. S. Life-Saving Service on the Pacific Coast*, Bandon Historical Society Press, Bandon, OR.

Peterson, Emil R. and Alfred Powers

1952 *A Century of Coos and Curry*, Binford and Mort, Portland, OR.

Sabel, Lynne and Phillip Steel

1977 *1,000 Great Events*, Exeter Books, New York, NY.

Sengstacken, Agness Ruth

1942 *Destination, West!*, Binford and Mort, Portland, OR.

West, Victor

1986 *Southern Oregon Mosquito Fleet-Stories About Coos County Boats 1852- 1948,* Bandon Historical Society, Bandon, OR.

Woolridge, Alice H.

1971 *Pioneers and Incidents of the Upper Coquille Valley, 1890-1940,* The Mail Printers, Myrtle Creek, Oregon.

INDEX